DIAMONDS AND CONFLICT:
PROBLEMS AND SOLUTIONS

DIAMONDS AND CONFLICT: PROBLEMS AND SOLUTIONS

ARTHUR V. LEVY (EDITOR)

Novinka Books
New York

Senior Editors: Susan Boriotti and Donna Dennis
Coordinating Editor: Tatiana Shohov
Office Manager: Annette Hellinger
Graphics: Wanda Serrano
Editorial Production: Marius Andronie, Maya Columbus, Vladimir Klestov, Matthew Kozlowski and Tom Moceri
Circulation: Ave Maria Gonzalez, Vera Popovic, Luis Aviles, Raymond Davis, Melissa Diaz, Magdalena Nunez, Marlene Nunez and Jeannie Pappas
Communications and Acquisitions: Serge P. Shohov
Marketing: Cathy DeGregory

Library of Congress Cataloging-in-Publication Data
Available Upon Request

ISBN: 1-59033-715-8.

Copyright © 2003 by Novinka Books, An Imprint of
Nova Science Publishers, Inc.
400 Oser Ave, Suite 1600
Hauppauge, New York 11788-3619
Tele. 631-231-7269   Fax 631-231-8175
e-mail: Novascience@earthlink.net
Web Site: http://www.novapublishers.com

All rights reserved. No part of this book may be reproduced, stored in a retrieval system or transmitted in any form or by any means: electronic, electrostatic, magnetic, tape, mechanical photocopying, recording or otherwise without permission from the publishers.

The authors and publisher have taken care in preparation of this book, but make no expressed or implied warranty of any kind and assume no responsibility for any errors or omissions. No liability is assumed for incidental or consequential damages in connection with or arising out of information contained in this book.

This publication is designed to provide accurate and authoritative information with regard to the subject matter covered herein. It is sold with the clear understanding that the publisher is not engaged in rendering legal or any other professional services. If legal or any other expert assistance is required, the services of a competent person should be sought. FROM A DECLARATION OF PARTICIPANTS JOINTLY ADOPTED BY A COMMITTEE OF THE AMERICAN BAR ASSOCIATION AND A COMMITTEE OF PUBLISHERS.

Printed in the United States of America

CONTENTS

Preface		vii
Chapter 1	Diamonds and Conflict: Policy Proposals and Background *Nicholas Cook*	1
Chapter 2	Diamond-Related African Conflicts: A Fact Sheet *Nicholas Cook and Jessica Merrow*	59
Chapter 3	Diamonds of Death *Ken Silverstein*	65
Chapter 4	Conflict Diamonds: Sanctions and War *Anna Frangipani Campino*	73
Chapter 5	International Trade: Significant Challenges Remain in Deterring Trade in Conflict Diamonds *Loren Yager*	81
Chapter 6	U.S. Initiatives on "Conflict Diamonds" *Office of the Spokesman Director*	99
Chapter 7	Sierra Leone: "Conflict" Diamonds Progress Report on Diamond Policy and Development Program *United States Agency for International Development (USAID) and Office of Transition Initiatives (OTI)*	103
Index		141

PREFACE

The mining and sales of diamonds by parties to participants of armed conflicts, notably in Africa, are regarded as a significant factor fueling such hostilities. These diamonds, labeled "conflict diamonds," make up an estimated 3.7% to 15% of the value of the global diamond trade. In response to public pressure to halt the trade of conflict diamonds and due to the persistence of diamond-related conflicts, governments and multi-lateral organizations have taken diplomatic action to combat the trade. Several international policy making forums, including the UN, have addressed the problem, and the international diamond industry and nongovernmental organizations have proposed a range of reforms and legislative initiatives to halt the illicit trade. This new book examines the diamond-related African conflicts, the various methods that are being researched and used to "tag" diamonds with microscopic markings, and the efforts taken to regulate the marketing and exporting of diamonds.

Chapter 1

DIAMONDS AND CONFLICT: POLICY PROPOSALS AND BACKGROUND

Nicholas Cook

INTRODUCTION

The Clinton Administration worked to create an international diamond trade regime, likely employing certificates of origin, but sought to ensure that legitimate diamond producers are not hurt by emerging policies. The Clinton Administration encouraged marketing reform and regulatory capacity building in African diamond producing countries. It consulted with the diamond industry, used U.S. membership on the United Nations (U.N.) Security Council to push for international sanctions banning the conflict diamond trade, and created an inter-agency group of conflict diamonds. The United States is participating in multi-lateral policy coordination initiatives and has sponsored and participated in a variety of conflict diamond policy-making forums. Critics of the Clinton Administration called for faster U.S. action to halt the conflict diamond trade. The Bush Administration has pursued policies on conflict diamonds that broadly mirror those of the Clinton Administration.

CONFLICT DIAMONDS: BACKGROUND AND POLICY

Recent Developments

The Kimberley Process, and inter-governmental forum that is crafting a proposal to create a regime to regulate international trade in diamonds, met in late November 2001 in Gaborone, Botswana. The Gaborone meeting sought to finalize detailed proposals for an international rough diamond certification scheme. On October 10, 2001, the Subcommittee on Trade of the House Committee on Ways and Means held a hearing entitled "Conflict Diamonds." The hearing assessed the status of the Kimberley Process, progress made in stopping the conflict diamond trade, and prospects for the passage of current legislation on conflict diamonds. Conflict diamond-related bills introduced in the 107th Congress include H.R. 918 (Hall); H.R. 2722 (Houghton); H.R. 2506 (Kolbe); S. 787 (Gregg); and s. 1084 (Durbin). H.R. 2722 (Houghton) passed the House by 408-6 on November 29 and was received in the Senate, where it was read twice and placed on the Senate Legislative Calendar under General Orders, Calendar No. 248. S. 1215 (Hollings), the Senate version of the Appropriations bill FY2002, Commerce, Justice, State would have enacted S. 787 into law, but was indefinitely postponed in the Senate by unanimous consent; diamond-related language in H.R. 2500 (Wolf), which became law and was the House alternative to S. 1215, was dropped during conference.

Introduction: Key Issues

The mining and sales of diamonds by parties involved in armed conflicts, notably in Africa, are regarded as a significant factor fueling such hostilities. These diamonds have been labeled "conflict diamonds" or "blood diamonds" because they fund purchases of arms and military material by belligerent forces. Trade in diamonds is a contributing factor to conflicts in Angola and the Democratic Republic of Congo (DRC). In Sierra Leone, where a fragile peace is taking hold, contention over control of diamond resources continues to generate political tensions and could potentially lead to renewed armed conflict. Diamonds have also contributed to the internationalization of all three civil conflicts. The possibility of gaining access to diamond wealth appears to have motivated foreign actors – including governments, private security-cum-mining firms, and mercenaries

– to become party to each of these conflicts, reportedly in exchange for diamond mining rights.

Although most observers view the causes of these conflicts as complex, reflecting a combination of political and socio-economic factors, analysts have increasingly focused on the connection between contested natural resources and political conflict.[1] The World Bank, for instance, published a study on causal connections between natural resources, demographic characteristics, and the occurrence of conflict. The Bank paper portrays diamonds as a particularly concentrated example of what it terms "lootable commodities," which its analysis indicates is an important factor driving conflict.[2]

World Diamond Market

World-wide diamond mine production in 2000 was estimated to be worth $7.86 billion, up from between $6.857 and $7.25 billion in 1999. In 2000, a total of nearly $9 billion of rough diamonds came to market globally, of which $5.67 billion was reportedly sold by the De Beers Diamond Trading Company (DTC, formerly called the Central Selling Organization).[3]

US Diamond Imports

U.S. diamond market demand is the largest in the world, and the vast majority of diamonds sold in the United States are imported. These imports bolster a large U.S. diamond retail jewelry market, which in 2000 was worth an estimated $26 billion – an increase of about 6% over 1999 – and comprised about 48% of the global $57.5 diamond jewelry retail market, which had grown from an estimated $56 billion in 1999.[4]

---

[1] Michael T. Klare. 2001. *Resource Wars: The New Landscape of Global Conflict*, 1st ed. New York: Metropolitan Books, *inter alia*.
[2] See the World Bank project, headed by economist Paul Collier, titled *The Economics of Civil War, Crime, and Violence*, which is available online. [http://www.worldbank.org/research/conflict/].
[3] Luc Rombouts. "Diamonds." Mining Annual Review. October 2001.
[4] The Diamond Registry, "2000 U.S. Diamond Jewelry Retail Sales Increase – May 2001," [http://www.diamondregistry.com/News/2001/retail.htm]. Other retail market estimates range from $27.6 to $30 billion. Andrew Coxon, De Beers LV; and Holly Burkhalter, "Blood On the Diamonds," Washington Post, November 6, 2001.

Table 1. Diamond Imports to the United States
($ millions)

Year	All Unmounted Diamonds*	Diamonds Cut, .5 carat or over**	Rough Diamonds***
1996	6,625	3,848	731
1997	7,635	4,719	646
1998	8,522	5,443	588
1999	9,929	6,321	733
2000	12,093	8,137	741

*Harmonized Tariff Schedule (HTS) – 7102: Diamonds, whether or not worked, but not mounted or set
**HTS – 7102.39.0050: Diamonds, non-industrial, (cut, faceted, set or mounted) 0.5 carat or over
***HTS – 7102.31: Diamonds, non-industrial, unworked or simply sawn, cleaved or bruted. The provisional definition of rough diamonds in the Kimberley Process Working document (see below) contains additional HTS categories.
Source: Interactive Tariff and Trade DataWeb (2001). U.S. International Trade Commission. Department of Commerce. [http://dataweb.usitc.gov/].

Conflict Diamonds and Production Statistics

The value of diamond production in 2000 in the three main countries from where conflict diamonds primarily originate has been estimated as follows: Angola, $739.7 million; D.R. Congo, $585 million; and Sierra Leone, $87.5 million.[5] Of that production, only a fraction is comprised of conflict diamonds, but that proportion is not definitively known. Estimates of annual diamond production and trade value in countries where conflict diamonds are mined vary widely because they rely on disparate methodologies and because detailed data are often lacking. Some estimates, for instance, take into account only official production figures, which ordinarily reflect official sales and production, and some do not take into account artisanal and unofficial production.[6] As a sub-portion of unofficial trade and production, conflict diamond trade and production volumes are especially difficult to measure but are believed to be significant. De Beers, a large diamond mining and marketing business group, has estimated that conflict diamonds comprised approximately 3.7% of world diamond

---

[5] Rombouts. "Diamonds."
[6] Some estimates, nevertheless, attempt to account for unofficial trade and production, using various proxy measurements, such as relative increases in exports from regions bordering production countries, field reports of artisanal production and small-scale trade, and confidential information from traders in international diamond processing centers.

production in 1999.[7] Other estimates range as high as 15% of the world trade in recent years. Some analysts, however, dispute the latter figure, asserting that it includes illicitly traded diamonds that do not have an origin in conflict. The illicit nature of conflict diamonds makes it extremely difficult to differentiate these gems from other illicit stones, according to many observers, because both varieties are traded in a similarly illegal fashion.[8]

CONTROLLING CONFLICT DIAMONDS

Policy Background

Nongovernmental organizations (NGOs) working on such issues as natural resource exploitation, human rights, and conflict resolution, have undertaken international advocacy campaigns aimed at halting the conflict diamond trade. A number of NGOs have formed a joint project called the *Fatal Transactions International Diamond Campaign*.[9] As public knowledge of the problem has grown, and in response to the persistence of diamond-related conflicts, international governmental organizations (IGOs) and national governments have undertaken highly publicized legal, diplomatic, and military actions aimed at combating the trade in conflict diamonds. Among the IGOs that have acted to address the problem are the United Nations (U.N.), European Union (EU), Economic Community of West African States (ECOWAS), and Southern African Development Community (SADC). Several recent multi-lateral conferences held in South Africa have focused on solutions to the conflict diamond problem (see *The Kimberley Process*, below).

The release of the *Fowler Report*, which described the status of the implementation of U.N. sanctions against the Angolan rebel group, UNITA, including a ban on the export and sale of Angolan conflict diamonds, motivated widespread concern and recognition of the connection between

---

[7] U.S. Congress. House. Committee on International Relations. Subcommittee on Africa. *Africa's Diamonds: Precious, Perilous Too?* Hearing, May 9, 2000, 106th Congress, 2nd session. Washington, U.S. Govt. Print. Off., 2000. Serial No. 106-142, page 102.

[8] The December 2000 report of the United Nations sanctions committee on Sierra Leone contains extensive discussions of disparities between production, trade, and conflict diamond statistics. See, particularly, paragraphs 112 to 150. United Nations. Security Council, *Report of the Panel of Experts appointed pursuant to Security Council resolution 1306 (2000)*, paragraph 19, in relation to Sierra Leone. S/2000/1195. December 20, 2000. [http://www.un.org/Docs/committees/SLTemplate.htm].

[9] The *Fatal Transactions Campaign* is online at the Netherlands Institute for Southern Africa Web site [http://www.niza.nl/uk/campaigns/diamonds/index.html].

conflict and the illicit diamond trade. The Fowler Report is the informal title of the March 2000 *Report of the Panel of Experts on Violations of Security Council Sanctions Against UNITA* (United Natons S/2000/203) and is named after the chairman of the sanctions committee, then-Canadian Ambassador to the U.N. Robert R. Fowler.[10] Several UN Security Council sanctions committee reports have since focused on the conflict diamond trade.[11]

The African Diamond Trade and Links to Terrorist Groups

Illicit trading in gems has recently been tied to the terrorist attacks of September 11, 2001. Several recent press accounts and testimony during the 2001 trial of several operatives of the Osama bin Laden's al Qaeda terrorist network have linked trade in conflict diamonds and other gems with the financing of al Qaeda. Testimony during the trial of four defendants convicted of participating in the bombing of the U.S. embassies in Kenya and Tanzania in August 1998 described trading in diamonds, tanzanite, rubies, and sapphires during the mid-1990s by bin Laden business associates.[12] In testimony before the Subcommittee on Trade of the House Committee on Ways and Means October 10, 2001 hearing entitled "Conflict Diamonds," Senator DeWine, Representative Hall, and Representative Wolf related efforts to combat trade in conflict diamonds to the U.S. global campaign against terrorism.[13]

---

[10] See Appendix, below, for more details on the *Fowler Report* and diamonds in Angola.
[11] Most of the sanction committee reports are available online. See: [http://www.un.org/Docs/sc/committees/INTRO.htm].
[12] See FBI – New York Office – Press Release, November 4, 1998, [http://www.fbi.gov/contact/fo/nyfo/pressrels/1998/11041998.htm], which includes links to indictments of Usama Bin-Laden and his alleged associates. See also daily transcripts of court proceedings in the case of United States of America v. Usama Bin Laden, et al., S(7) 98 Cr. 1023, [http://cryptome.org/use-v-ubl-dt.htm]; and related court documents, e.g., "Complaint Violations of Title 18, United States Code, Section 1001," United States of America v. Wadih El Hage, a/k/a/ 'Abdus Sabbur,' Defendant. [http://www.ccc.de/mirrors/jya.com/usa-v-qaeda.htm]; and Judy Aita, "FBI Agent Recounts Confession of Bombing Trial Defendant," U.S. Department of State, Office of International Information Programs, March 1, 2001.
[13] U.S. Congress. House of Representatives, Committee on Ways and Means. Subcommittee on Trade, *Conflict Diamonds*, Hearing 107th Congress, 1st Session, October 10, 2001, [http://waysandmeans.house.gov/trade/107cong/tr-6wit.htm].

Alleged al Qaeda-RUF Relationship[14]

A November 2, 2001 *Washington Post* article, attributing its facts to "U.S. and European intelligence officials" alleges that "[d]iamond dealers working directly with men named by the FBI as key operatives in bin Laden's al Qaeda network bought," as well as Hizballah representatives, have purchased diamonds from members of the Sierra Leone Revolutionary United Front (RUF).[15] The Post account, by Douglas Farah, also ties sales of RUF diamonds funding to the southern Lebanese Hizballah militia movement, and notes that a minority of diamond traders in the Lebanese Diaspora in Africa have long been believed to be involved in such activities. Other press accounts have mirrored this assertion, and have tied similar activities in Angola to the funding of the Lebanese Amal militia.[16]

Liberian officials have called Farah's account inaccurate, but have not categorically denied it. U.S. State Department officials have stated that they do not have direct evidence corroborating Farah's account. RUF officials have denied having any links with al Qaeda or selling diamonds to the organization, but have reportedly acknowledged that such sales could have taken place without their knowledge.[17] Omrie Golley, an RUF official who chairs the Sierra Leone Political and Peace Council, reportedly stated that a panel would be established to investigate the reports.[18] Recent press reports suggest, however, that extensive diamond mining activities in the diamond-rich Eastern Province of Sierra Leone – where disarmament has not been completed and where the Sierra Leone police and UNAMSIL peacekeepers reportedly have only a skeletal presence – are occurring. Some observers

---

[14] The following account summarizes the *Washington Post*, "Al Qaeda Cash Tied to Diamond Trade," and draws from the report of the U.N. panel of experts monitoring implementation of U.N. Security Council Resolution 1343 (2001) concerning Liberia, S/2001/1015.

[15] See Douglas Farah, "Al Qaeda Cash Tied to Diamond Trade Sale of Gems From Sierra Leone Rebels Raised Millions, Sources Say," *Washington Post*, November 2, 2001, page A1.

[16] See Panafrican News Agency, "Belgium Accused Continuing Sale of UNITA Diamonds," April 24, 2001; and Agence France Presse, "Belgian diamond traders dealing with Angolan rebels: press," April 23, 2001. On Al Qaeda, see CRS Terrorism Briefing Book EBTER131, *Al Qaeda* and CRS Report RL30588, *Afghanistan: Current Issues and U.S. Policy Concerns*. On Hizballah, see CRS Report IB93033, *Iran: Current Developments and U.S. Policy: Issue Brief* and CRS Report RL30713, *The Current Palestinian Uprising: Al-Aqsa Intifadah*.

[17] The RUF, which has been named as a terrorist group by the U.S. State Department and is infamous for waging a brutal war against the government of Sierra Leone, has agreed to disarm and participate in scheduled elections. See U.S. State Department, *Patterns of Global Terrorism 2000* [http://www.state.gov/s/ct/rls/pgtrpt/2000]. See also CRS Report RL30751 *Diamonds and Conflict: Policy Proposals and Background* and CRS Report RL31062 *Sierra Leone: a Tentative Peace?*

[18] See IRIN News, "Sierra Leone: RUF denies terrorist links," November 5, 2001 and VOA News, "Reports Say Sierra Leone Diamonds Funding Al-Qaeda," November 3, 2001.

have questioned where newly mined diamonds are being sold, by whom are they being sold, and to whom such profits are accruing. Farah's account suggests that this heightened activity may be in direct response to requests by al Qaeda buyers, although it also implies that RUF leaders may not have known the identity or affiliation of the al Qaeda buyers.

Al Qaeda Diamond Purchasing Operation

Ibrahim Bah, a key RUF official who was a subject of a recent report by a panel of experts investigating compliance with U.N. Security Council resolution 1343 (2001) concerning Liberia, S/2001/1015, is reported by Farah to have primary responsibilities for marketing RUF diamonds. He is also reported to be a key middleman in transactions between the RUF and al Qaeda and Hizballah diamond buyers. According to Farah, Bah is a former member of the separatist Movement of Democratic Forces of Casamance (MFDC) of Senegal, his birthplace. He is said to have later trained in Libya under the patronage of Libyan leader Muammar al-Qadhafi and to have fought as a "majahedin" in Afghanistan against Soviet forces in the early 1980s, and later in the Hizballah militia against Israel. Bah, who is said to live in Burkina Faso, later returned to Libya, and was allegedly involved in training Liberian President Charles Taylor, a former civil war faction leader, and RUF founder Foday Sankoh.[19] Bah reportedly fought in Liberia with the forces of both men. Taylor, whose government faces U.N. Security Council sanctions related to its reported involvement in trading smuggled Sierra Leone RUF diamonds, has repeatedly denied his involvement in such dealings.

Beginning in September 1998, Farah reports, al Qaeda buyers were able to purchase diamonds at sub-market prices and sell them in Europe at a steep profit worth several – perhaps tens – of millions of dollars.[20] Since mid-2001, however, the account asserts that top prices have been paid by the al Qaeda operatives, possibly with the expectation that the organization's business or bank accounts could be frozen following the September 11 attack. The heightened transaction activity could also indicate an increased need to launder funds from a variety of sources – possibly including other illicit dealings, such as drug sales – and to transfer their value into the

---

[19] Burkinabe president Blaise Compaore allegedly also has close ties with the al-Qadhafi and Taylor regimes and has been linked in U.N. and press reports to RUF diamond trading.
[20] No precise figure is known. The volume of RUF trade, generally, is difficult to ascertain, because no independent monitoring of diamond mining in areas of eastern Sierra Leone control by the RUF has been undertaken, and because RUF sales are, by their nature, covert.

fungible, concentrated form of value that diamonds represent. Diamonds are said to have been smuggled into Liberia from Sierra Leone by "senior RUF commanders," and are then said to have been purchased by al Qaeda-affiliated dealers in transactions supervised by Bah. The dealers flew between Belgium and Monrovia several times monthly to make purchases, which are reportedly undertaken in premises safeguarded by Liberian government security forces. In Monrovia the dealers "are escorted by special [Liberian] state security through customs and immigration control."

Farah's reports that the RUF association with al Qaeda began when Abdullah Ahmed Abdullah, who according to the FBI is "indicated for his alleged involvement in the August 7, 1998, bombings of the United States Embassies in Dar es Salaam, Tanzania, and Nairobi, Kenya, "made an initial trip to Liberia in September 1998.[21] He is aid to have met with Sam "Mosquito" Bockarie – a former RUF commander who was resident in Liberia and who is reported to have close relations to Charles Taylor – to arrange future diamond transactions. The visit was then reportedly followed by a diamond buying trip by tow other alleged al Qaeda operatives, Ahmed Khalfan Ghailani and Fazul Abdullah Mohammed. Both have been indicted in the United States for their alleged involvement in the Tanzania Kenya U.S. embassy bombings.[22] Later al Qaeda diamond transactions were reportedly undertaken by Lebanese diamond dealers, two of whom are identified by the Post as Aziz Nassur and Sammy Ossailly, both reportedly based in Belgium. Ghailani and Fazul are also said to have later made field visits to RUF-held areas in eastern Sierra Leone, but were said to be noticeable to locals as outsiders, and were replaced by Senegalese buyers based in a Monrovia safe house.

Al Qaeda and the Tanzanian Gem Trade

On November 6, 2001, the *Wall Street Journal* reported that the Tanzanian government is investigating illicit trading and a tanzanite smuggling network with alleged links to Osama bin Laden's al Qaeda network.[23] Tanzanite, a gemstone that turns blue when heated, is rare, having been discovered at a single, small site in Mererani, Tanzania. The *Wall Street Journal* account describes the rise in Mererani of a radical,

---

[21] See Federal Bureau of Investigation, *Most Wanted Terrorists* [http://www.fbi.gov/mostwant/terrorists/fugitives.htm].

[22] *Ibid.*

[23] Robert Block and Daniel Pearl, "Much-Smuggled Gem Called Tanzanite Helps Bin Laden Supporters," *Wall Street Journal*, November 16, 2001, page A1, A8. A similar account appeared in Africa Confidential, "Gems for the Martyrs," 42:23, November 23, 2001, page 3.

fundamentalist Islamic group – one of several in Tanzania, a country where tolerant, moderate forms of Islam predominate – centered around an imam known as Sheik Omari. According to the account, Omari recently opened a mosque, Taqwa, and urges his followers – many of whom are reportedly active in the tanzanite trade – to use their commercial activities to promote Islamic militancy. The gems are described as being illicitly exported by associates of Omari to Dubai, which has been identified by U.S. investigators as a key operational locus of al Qaeda financial dealings, and to Hong Kong.[24]

Conflict Diamonds: Recent Public Debate

Conflict diamonds received extensive coverage in the U.S. and international press, and conflict diamonds gained a much-heightened profile in popular U.S. electronic media grew in 2000 and 2001. Several U.S. network TV news magazines and evening new shows, and at least one prime time TV drama, covered the issue.[25]

Concern over increased negative publicity about conflict diamonds grew among some in the diamond industry. Some governments, both in diamond producing and consuming nations, and several major firms and marketing organizations worried that an uninformed public, unable to distinguish conflict diamonds from legitimate diamonds, might begin to associate all diamonds with conflict and decrease their purchases as a result. Such a trend, it was feared, might undermine the wholesale and retail industries – but also the socio-economic development of stable and prosperous democratic African states such as South Africa, Botswana and Namibia, to which the legitimate production of diamonds contributes substantially.

---

[24] See, for instance, Edward Alden and Mark Turner, "US freezes more of bin Laden's financing: Bush issues new blacklist headed by two groups said to be main funders of al-Qaeda terror network" "Financial Times (London), November 8, 2001; David S. Hilzenrath and John Mintz, "European Bank Regulators Help Track al Qaeda Assets; Reports Solicited on Contact With Banks Tied to Bin Laden," Washington Post, September 29, 2001, page A19; Glenn R. Simpson, "U.S. Intensified Financial War On Terrorists," Wall Street Journal, November 8, 2001, page A3; Warren Hoge, "In Emirates, An Effort To Examine Bank System," New York Times, October 15, 2001, page B6; and Agence France Presse, "US Says Gulf Bank Laundered Money For Bin Laden," July 8, 1999.

[25] See, for instance, Dateline NBC Investigation, "Diamonds in Conflict," NBC News, July 1, 2001, [http://www.msnbc.com/news/593785.asp]; Bob Simon, "Diamonds: A War's Best Friend," CBS News, June 14, 2001; Law & Order, "Soldier of Fortune," NBC.com Episode Guide, October 24, 2001; and John Martin, "Dirty Diamonds Dilemma," ABCNews.com, ND.

The views of U.S. officials have reflected such concerns; Bush and Clinton Administration officials have, respectively, expressed support for efforts to sharply distinguish legitimately produced diamonds from conflict diamonds. During an April 2001 visit to Botswana, a U.S. Congressional delegation examined the role played by diamonds in the country's development and in the funding of HIV/AIDS treatment (see below).[26]

Advocacy Campaigns

The majority of NGOs advocating increased regulation of the diamond trade agree that the great majority of diamonds are legitimately produced and generate crucial socio-economic benefits, and rejected a blanket diamond consumer boycott of diamonds.[27] Many activists have, however, urged consumers to boycott conflict diamonds and to assess the ethics of purchasing diamonds that could not be independently verified as being conflict-free. In 2000 and 2001, activists mounted publicity campaigns and limited numbers of demonstrations, picking major diamond retailers. They mounted highly emotional campaigns, arguing that diamond jewelry purchased to express sentiments of love may have been produced by persons subject to severe human rights abuses or sold by persons committing such abuses. In congressional hearings, press conferences, and in TV and online commercials, activists used graphic images of explicitly linked and contrast amputations of limbs and social disintegration – phenomena associated with conflict diamonds – with diamonds and their widely-held association with love and the social union of marriage.

Producer Reactions to Market Threats

To counter the threat posed by possible consumer rejection of diamonds, some diamond producing countries and industry trade groups undertook public education and legislative lobbying campaigns. They sought to ensure that the legitimate diamond industry was not tarnished by conflict diamonds and endeavored to influence the passage of conflict diamond-related

---

[26] See Bruce Alpert and Bill Walsh, "On the Hill; News from the Louisiana Delegation in the Nation's Capital," The Times-Picayune (New Orleans), April 8, 2001, page 8; BBC Worldwide Monitoring, "Botswana: US Congress Delegation Pronounces Botswana Diamonds "Conflict Free," text report of Radio Botswana, Gaborone, April 12, 2001; and South African Press Association, "US Congressmen Check on Botswana's 'Clear' Diamonds," April 9, 2001.

[27] A minority of activists, however, have taken the position that if diamond trading reforms are not undertaken rapidly, the diamond consuming public might be convinced to undertake a general anti-diamond buyers boycott akin to the economically significant consumer boycott of fur in the 1980s and 1990s.

legislation that would not restrict or decrease trade in legitimate diamonds.[28] Such efforts included the following:

- **Debswana.** In March 2001, Debswana, a diamond producing firm owned in equal share by the Botswana Government and De Beers, reportedly hired the lobbying firm Hill and Knowlton to influence conflict diamond-related legislation and to undertake public affairs programming promoting the positive role played by diamonds.[29]

- **World Diamond Council.** In 2000, the World Diamond Council (see below) published a website outlining its contributions to policymaking, legislation, and public debate on conflict diamonds. The WDC has been an active in the Kimberley Process (see below).

- **Jewelers of America.** The Jewelers of America (JA) has actively countered possible negative effects of consumer perceptions of diamonds as a result of publicity about the conflict diamond trade, and has contributed to the formulation of policies to end it. Matthew Runci, JA president and CEO, has testified in Congressional hearings several times about his group's efforts to end trade in conflict diamonds, and has contributed to the Kimberley Process meetings. JA has urged its members "to the best of your ability...[to] undertake reasonable measures to help prevent the sale of illicit diamonds" while acknowledging that "it is not currently possible for retail jewelers to verify the country of origin of diamonds."[30]

Botswana

Debswana's hiring of Hill and Knowlton was reportedly part of a public diplomacy campaign by the Botswana government entitled Diamonds for Development. Diamonds account for about 79% of Botswana's total export earnings, just over 40% of its gross domestic product, and reportedly over

---

[28] Greg Mills, "From Conflict to Prosperity Diamonds?: The Role of Diamonds as a Development Asset in Africa," international Ministerial Diamond Conference, [http://www.24hourdiamondnews.com/gov.htm].

[29] Political Finance & Lobby Reported, "Lobby registrations: Lawyers & consultants: international Trade," March 28, 2001; Bruce Alpert and Bill Walsh, "Headline: On The Hill; News from the Louisiana delegation in the nation's capital," The Times-Picayune (New Orleans), April 8, 2001, page 8.

[30] Jewelers of America, "JA Takes Initiative on African Diamond Controversy," June 19, 2000.

half of government revenues.[31] The campaign focuses on the key role of "prosperity diamonds" in funding Botswana's socio-economic and economic successes, and in bolstering the country's democratic and civil conflict-free record. It has included overseas trips by top Botswana officials, including a June 2001 visit to the United States by Botswana's President, Festus Mogae.[32] It highlights Botswana's high rates of literacy, access to clean water and health care, high per capita income levels, its extensive transport and communications infrastructure, and the government's provision of drug treatment for the large (38.5% of adults in 2000) HIV/AIDS-afflicted population.[33]

Regulatory Challenges

Effective policing of the illicit diamond trade faces difficult challenges. Diamonds are a highly fungible and concentrated form of wealth, and the legitimate international diamond industry is historically insular, self-regulating, and lacks transparency. The trade in conflict diamonds takes advantage of these factors. The conflict diamond trade has also been linked to covert and sometimes violent business transactions. It is reportedly associated with international criminal activities, such as money laundering, smuggling, commercial fraud, and arms trafficking. Observers have

---

[31] Data derived from World Development Indicators Database and World Bank, "Botswana at a Glance," September 27, 2001; and SDI Magazine, "Why gems must cover the cost of progress," May/June, 2001.

[32] See, for example, Kathy Chenault, "Diamonds and Dreams: Botswana Worries that Stigma of Gems' Role in African Wars Could Undermine Its Progress," San Francisco Chronicle, June 4, 2001; Debswana, "Development Diamonds," [http://www.debswana.com]; Botswana High Commission [London], "Botswana Diamonds for Development," [http://www.diamondsfordevelopment.com].

[33] On socio-economic data, see World Development Indicators Database and "Botswana at a Glance"; and World Health Organization, "Botswana 2000 Update (revised)," Epidemiological Fact Sheet of HIV/AIDS and Sexually Transmitted Infections. Despite its democratic record, the Botswana government has been criticized for instituting policies that allegedly threaten the cultural and economic survival of the traditionally nomadic San (Bushmen) people. Such policies reportedly include lack of recognition for traditional San land use rights and hunting restrictions in areas where diamonds have been discovered – such as Gope in the Central Kalahari desert. Such policies allegedly compel the San, for economic reasons, to accept government welfare and communal relocation to sites – seen as marginal lands by critics – designated by the government. Some observers, while acknowledging the important role diamonds have played in Botswana's development, believe that the country must diversify if it is to maintain a healthy economy in the long term. See, inter alia, "Botswana's treatment of San minority provokes demonstrations," afrol News, June 27, 2001.

concluded that conflict diamonds regularly enter the legitimate international market.[34]

Policy Proposals[35]

Most proposals for curtailing the trade in conflict diamonds center around implementing systems to identify the origin of diamonds to ensure that diamonds sold by illicit sellers do not enter legitimate international commerce.[36] Such proposals provide the basis for laws and international actions, such as U.N. Security Council sanctions, that ban trade in conflict diamonds. Three primary approaches for determining the provenance of diamonds have been proposed.

1. Physical or "Geo-Chemical" Identification of Diamonds

Research on geo-chemical methods for diamond identification focuses on the comparative analysis of trace elements and impurities within diamonds. Such information would be used to establish common characteristics of diamonds from similar areas or to pinpoint unique characteristics – in a manner analogous to fingerprinting – of individual diamonds. This research employs plasma mass spectrometry and related technologies.

A related approach is to identify and correlate the surface, crystal, and other structure-related characteristics of right (unpolished) diamonds, with the aim of establishing unique identifying features of groups of diamonds from similar places of origin, and possibly of individual gems. Such identification would be based on visual assessments and on the use of

---

[34] Ian Smillie, Lansana Gberie and Ralph Hazleton describe and cite a range of illegal and gray-market operations associated with diamond trading. See The Heart of the Matter: Sierra Leone, Diamonds and Human Security, Partnership Africa Canada, January 2000, online at [http://www.partnershipafricacanada.org/English/esierra.html]. Multiple press accounts also describe illicit acts associated with diamond trading.

[35] Comprehensive treatment of technical and policy issues related to conflict diamonds is contained in Global Witness, *Conflict Diamonds: Possibilities for the Identification, Certification and Control of Diamonds*, May 2000, which is also available online at [http://www.oneworld.org/globalwitness/reports/conflict/cover.html]. Also see statement of William E. Boyajian, President, Gemological Institute of America, and on behalf of the World Diamond Council, *Testimony Before the Subcommittee on Trade of the House Committee on Ways and Means Hearing on Trade in African Diamonds*, September 13, 2000. Online at [http://waysandmeans.house.gov/trade/106cong/9-13-00/9-13boya.htm].

[36] The *origin* of a diamond refers to its physical origin, or place where it was mined. A diamond's *provenance* refers to the place from where it was last shipped. In published accounts describing the diamond industry, the two terms have sometimes erroneously been conflated.

spectral refraction methods or optical, laser, x-ray, and other scanning technologies.

2. Tagging of Diamonds

This approach seeks to use laser and focused ion beam technologies to inscribe on individual diamonds identifying information, such as microscopic bar codes, which can then be used to register and track stones. Several firms market the requisite technology. Other firms offer technology that can be used to identify unique spectral features of individual cut diamonds using laser-scanning technologies. The costs of tagging technology currently represent a barrier to their widespread use in diamond commerce, but expert opinion suggests that these prices may fall in the near to medium future. Critics point out that it may possible to cut off or otherwise obliterate identifying marks that are cut onto the face of diamonds.

3. Certificate of Origin Laws

This approach seeks to create a legally binding chain of warranties from the point of mining origin to the country of importation or – in some proposals – to the retail level. The objective is to create trade documentation that, based upon verification by the authorities of an exporting country, validates the legal origin of diamonds. It forms the basis for findings of legal fact in efforts to track and monitor the diamond transactions. The approach relies on effective administrative processes and law enforcement; and proper adherence to prescribed regulatory procedures. The certificate of origin approach is likely to form an integral part of any future international diamond trade regime.

Industry Policy Initiatives

Diamond High Council

The Diamond High Council (HRD) is a formal trade organization representing the Belgian diamond industry. Antwerp, Belgium, where the HRD is headquartered, is one of the leading international diamond cutting centers, and is a major destination for exports of rough diamonds from Africa. The HRD has close working ties with the Belgian government. HRD has taken uni-lateral and multi-lateral action to curtail the trade in conflict diamonds. Beginning in late 1999, it assisted the Angolan government in designing a forgery-proof certificate of origin documentation system, and later entered into a joint export control regime and technical assistance agreement with the Angolan government. It has given the Sierra Leonean government similar certificate of origin advice, in coordination with donor governments, including the United States, and has stated that it will seek to establish a joint export control regime with Sierra Leone along the lines of the Angolan arrangement.

The HRD has stated that the Belgian Ministry of Economic Affairs, since February 2, 2000, has required that imports be licensed under the name of individual diamond dealers for all diamond imports from: Liberia, Ivory Coast; Uganda, Central African Republic; Ghana; Namibia; Congo (Brazzaville); Mali; and Zambia. No government-verified certificate of origin system currently exists in these countries, according to the HRD. The HRD has stated that if probable cause exists indicating that diamonds imported to Belgium do not originate in the country of export, Belgian government officials will attempt to determine the source of such stones. HRD has reportedly assisted Guinea and the DRC to set up diamond certification systems.[37]

World Diamond Council

In July 2000, during the World Diamond Congress in Antwerp, Belgium, the two largest international diamond trade organizations, the World Federation of Diamond Bourses (WFDB) and the International Diamond Manufacturers Association (IDMA), jointly issued a resolution calling for:

- A uniform, global export certification system, underpinned by national legislation in participating countries, establishing a range of export control mechanisms aimed at ensuring the legitimate origin of internally traded diamonds. Such legislation would require a system of seals and registration for the export of diamond parcels, controlled and maintained by national, internationally accredited export agencies; criminal penalties for illicit diamond trading; and a system for monitoring compliance with the system.

- The mandatory establishment by diamond trade organizations of ethical codes of business practice aimed at ensuring transparency and adherence to legal requirements in diamond commerce; and

---

[37] HRD, "Guinea First Country not in Conflict to Adopt Certification Scheme," May 2, 2001; and HRD, "D.R. Congo to Set Up Certification Scheme for Diamonds," April 27, 2001. [http://www.conflictdiamonds.com/pages/interface/newsframe.html].

cooperation in monitoring compliance with such codes and germane trade law.

Acting under the Antwerp Resolution, which called for the creation of the World Diamond Council (WDC), the WFDB and IDMA chartered this organization. In September 2000 in Tel Aviv, Israel, the WDC held an inaugural policy-planning meeting. According to testimony by Matthew A. Runci, President and Chief Executive Officer of the Jewelers of America, Inc., speaking on behalf of World Diamond Council before the House Committee on Ways and Means Subcommittee on Trade hearing on *Trade in African Diamonds*, September 13, 2000, [38] the WDC plan includes the following elements:

- Establishment of dedicated rough diamonds import/export offices that are closely supervised by individual government authorities;
- Adoption of a uniform international certification system requiring that all rough diamond parcels being traded internationally be sealed and authenticated prior to export;
- Monitoring of industry-wide compliance with proposed ethical codes of conduct that prohibit the trade in conflict diamonds;
- Obliging banks, insurance companies, shipping companies and other providers of auxiliary goods and services to cease business relations with any company or individual knowingly involved in dealing in conflict diamonds;
- The result of these steps will be to support a chain of assurance for traders of polished diamonds based on rough controls.

The WDC stated that these steps together will create a support chain of assured legitimacy of provenance for diamond traders. The WDC has also called upon governments of diamond exporting and importing countries to enact legislation that would support the WDC's goals, and has advocate the incorporation of its proposals in to the certification system being crafted by the Kimberley Process.[39] In November 2000, the WDC reportedly hired a

---

[38] Outline at [http://waysandmeans.house.gov/trade/106cong/9-13-00/9-13runc.htm].
[39] According to the WDC website, as of November 20, 2001, Sierra Leone and the DRC, two of the countries most affected by conflict diamonds are not represented on the WDC as Government Observers or Members. Angola is represented by Ascorp, a joint venture between the government and private shareholders that holds a monopoly on the sale of

law and lobbying firm, Akin, Gump, Strauss, Hauer & Feld, to draft model legislation on behalf of the WDC.[40] The model legislation, based on the WDC's proposed rough diamond export and import control system, was publicly released in January 2001.[41] The WDC has continued to be active in seeking to influence proposed congressional legislation in the 107th Congress.

De Beers

As of March 27, 2000, under the trademark initials DTC (for the Diamond Trading Company Limited, the gem-quality diamond sales arm of the De Beers group of companies), De Beers guarantees that it does not purchase or sell conflict diamonds (see above).[42]

DTC has also introduced formal rules for its 125 "sight" holders – or rough diamond wholesale buyers – replacing a reported system of informal, unwritten criteria with which sight holders were previously required to comply. The new system reportedly includes provisions requiring that sight buyers who are discovered to be purchasing diamonds not guaranteed as being "conflict-free" lose their right to purchase from De Beers, which reportedly controls sales of between 44 and 70% of the world rough diamond market. A De Beers representative has reportedly stated that its efforts and those of the industry at large have caused an approximate 30% price drop for conflict stones.[43]

Angolan diamond exports. See Members and Committees, World Diamond Council, [http://www.worlddiamondcouncil.com/memberscom.html].

[40] Judy Sarasohn, "$2 Million Assist Costs University Nothing," Washington Post, November 2, 2000, page A27; and Martin Rapaport, "WDC to Offer Model Statute to Curb Conflict Diamonds,"
[http://www.diamonds.net/news/newsitem.asp?num=4607&type=all&topic=all].

[41] See World Diamond Council, "A System for International Rough Diamond Export and Import Controls," [http://www.worlddiamondcouncil.com/system.html] and "U.S. Diamond Legislation Draft – December 23, 2000 – Bill Format," [http://www.worlddiamondcouncil.com/bill122300.html], and related WDC documents [http://www.worlddiamondcouncil.com].

[42] Some observers have raised doubts about the De Beers guarantees. See, inter alia, *Waiting on Empty Promises: The human cost of international inaction on Angolan sanctions*, by Action for Southern Africa, April 2000 [http://www.anc.org.za/Angola/actsareportv4.html].

[43] "WDC Outlines Action Plan," *The Mining Journal*, September 15, 2000, page 208; Sharad Mistry, "De Beers to Market Branded Diamonds as Competition Hots Up," *Financial Express*, May 29, 2000, which is also available online at: [http://www.financialexpress.com/fe/daily/20000529/fco29088.html]; Francesco Guerrera and Andrew Parker, "De Beers Seeks Curbs on Rebel Diamonds," *Financial Times*, July 7, 2000, page 1; and Francesco Guerrera and Andrew Parker, "De Beers: All that Glitters is Not Sold," *Financial Times*, July 7, 2000, page 12.

U.S. Policy

The efforts of the Clinton Administration to combat the trade in conflict diamonds focused on the creation of a multi-lateral diamond trade regime backed by international sanctions aimed at curtailing such commerce. Such a regime would have been based on a variety of formal working partnerships between legitimate diamond producing state; those that import, trade, and consume diamonds; the international diamond industry; and a range of non-governmental organizations. The Clinton Administration also sought to ensure that the industries of legitimate diamond producing African democratic states – particularly Namibia, Botswana, and South Africa – would not be harmed by efforts to curtail the trade in conflict diamonds.

International and Multilateral Policy

The Clinton Administration sponsored conferences focusing on the war economies of conflict diamond-producing states, and held unilateral policy dialogues with these and non-conflict producing states, such as Botswana. It also consulted with members of the American diamond industry. The Clinton Administration used U.S. membership on the U.N. Security Council to push for international sanctions banning the illicit trading of diamonds from Angola and Sierra Leone, and for the appointment of panels of experts to monitor compliance with these sanctions. The Security Council also appointed a panel of experts to examine the illicit exploitation of natural resources in the Congo. (See Appendix below). The Clinton Administration took unilateral actions to isolate and penalize governments that abet the trade in conflict diamonds or violate related U.N. resolutions. These included a October 10, 2000 presidential proclamation denying entry into the United States of persons who assist or profit from the armed activities of the Revolutionary United Front (RUF) rebels fighting the government of Sierra Leone. The restrictions applied to President Charles Taylor, senior members of the Liberian government, their supporters, and their families, and represented an explicit sanction against the Liberian government for its failure to end its trafficking in arms and illicit diamonds with the RUF, thus fueling the Sierra Leonean conflict.

The United States also participated in multi-lateral diplomatic and policy-focused coordination initiatives, both at the inter-governmental level, and in forums involving participation from governments of producing and consuming nations, NGOs, and the international diamond industry. One result of government-to-governments dialogue was a major policy statement in July 2000 by the Group of Eight (G8) on Illicit Trade in Diamonds.[44] U.S. efforts to encourage the July 2000 G8 joint statement were preceded by Secretary of State Madeleine Albright's December 1999 G8 Berlin Ministerial presentation, in which she highlighted the connection between arms and diamond trading.

Africa-Focused and Bilateral Policy

In addition to multi-lateral efforts, the Clinton Administration encouraged diamond marketing reform and the development of regulatory capacity in African diamond producing countries through unilateral dialogue and joint U.S.-African policy planning exercises. These efforts sought to assist African state to create sound legal and administrative mechanisms in order to better regulate their domestic diamond industries and to integrate these mechanisms with similar regulatory regimes in consuming and importing countries. The Office of Transition Initiatives of the Agency for International Development provided technical assistance to Sierra Leone – in partnership with other donor governments and industry officials – to develop an effective certificate of origin export system in Sierra Leone. It also encouraged increased transparency, competition, and participation-broadening reforms based on free market principles within Sierra Leone's domestic diamond industry.

Criticism of Clinton Administration Policy

In a statement before the House International Relations Committee Subcommittee on Africa during a May 9, 2000 hearing entitled *Africa's Diamonds: Precious, Perilous Too?,* Representative Wolf stated that "[w]hile the West lets the problem of conflict diamonds fester, conditions where this illicit trade occurs, continue to worsen...I have written to the Administration several times about the problems in Sierra Leone and about the issue of conflict diamonds...To date, the Administration has done little or nothing on any of these recommendations..."[45] During a September 13, 2000 hearing of the Trade Subcommittee of the House Ways and Means Committee entitled *Trade in African Diamonds,* several Members called for

---

[44] Ministry of Foreign Affairs of Japan, "G8 Miyazaki Initiatives for Conflict Prevention: 3 Illicit Trade in Diamonds," July 2000.
[45] "Statement by Frank R. Wolf," Testimony before the Subcommittee on Africa of the House International Relations Committee hearing on *Sierra Leone and Conflict Diamonds,* May 9, 2000 [http://www.house.gov/international_relations/af/diamond/wolf.htm].

more active Administration engagement to curtail the trade in conflict diamonds. Representative Hall stated that "there is apparently not the sustained commitment from senior [Clinton] Administration officials [that] this issue merits."[46] At the same hearing, Representative Cynthia McKinney stated that the United States "must show leadership and act more swiftly against all the countries mentioned in the *Fowler Report*."[47]

Clinton Administration Response

The Clinton Administration countered that it had actively worked to curtail the conflict diamond trade, but noted that international consensus on how to halt the trade in conflict diamonds – which it saw as a prerequisite for successful policymaking – had not emerged. Testifying before the House Ways and Means Subcommittee on Trade on September 13, 2000, William Wood, Principal Deputy Assistant Secretary of State for International Organization Affairs, cited Clinton Administration U.S. participation in the Kimberley Process and other policy forums, such as the G8. He noted that since 1998 the Clinton Administration had supported U.N. sanctions to prevent the trade in conflict diamonds, and described U.S. efforts to assist Sierra Leone and Angola to improve their diamond export certification systems. He welcomed legislation expressing a sense of the Congress in support of administration efforts to curtail the conflict diamond trade, but cautioned against legislation that would mandate specific policies which, he stated, might not conform with on-going efforts to craft an international diamond trade regime.

Clinton Administration officials also highlighted their support for Resolution 56 of the 55th Session of the U.N. General Assembly.[48] On January 10, 2001, the White House Office of Science and Technology Assessment, in conjunction with the National Security Council, the State Department, the National Science Foundation, and the Treasury Department, held a White House Diamond Conference entitled *Technologies for*

---

[46] "Statement of the Honorable Tony P. Hall, M.C., Ohio," Testimony before the Subcommittee on Trade of the House Committee on Ways and Means, Hearing on Trade in African Diamonds, September 13, 2000. Online at [http://waysandmeans.house.gov/trade/106cong/9-13-00/9-13hall.htm].

[47] See Appendix, below, for more details on the Fowler Report and diamonds in Angola. See "Statement of the Honorable Cynthia McKinney, M.C., Georgia," Testimony before the Subcommittee on Trade of the House Committee on Ways and Means Hearing on Trade in African Diamonds, September 13, 2000, online at: [http://waysandmeans.house.gov/trade/106cong/9-13-00/9-13mcki.htm].

[48] A/RES/55/56, [http://www.un.org/documents/ga/res/55/a55r056.pdf].

Identification and Certification. Nearly one hundred and fifty policy makers, scientists, engineers, and representatives of the world diamond industry and NGOs participated in the forum. They assessed the technical methods of determining the origin of rough diamonds; technologies to support an origin certification regime; and associated policy issues.

Bush Administration Policy on Conflict Diamonds

The Bush Administration appears, particularly in reference to Sierra Leone, to be pursuing policies to stem the flow of conflict diamonds that are broadly similar to those of the Clinton Administration. On January 25, 2001, in a statement to the U.N. Security Council during a review of the Panel of Experts Report on Sierra Leone Diamonds and Arms, Acting U.S. Representative to the U.N., Ambassador James B. Cunningham, stated that:

> Controlling the flow of conflict diamonds and illicit arms is essential to end the fighting and destabilization in Sierra Leone and its neighbors. We are intent on ending the illicit trade in arms-for-diamonds that has caused so much devastation and human suffering in Sierra Leone and throughout West Africa. We welcome the upcoming visit of ECOWAS ministers. We will work hard with Council Members, the UN and countries in the region to bring panel recommendations into begin and to deal firmly with illegal trade and with sanctions violators.[49]

During the January 25 Security Council session, representatives of the United States and Great Britain reportedly stated that they are working on a resolution that would impose mandatory worldwide sanctions in Liberia until it ceases its illicit diamonds-for-arms trafficking. The draft resolution calls for a global prohibition on the direct or indirect import of all rough diamonds from or through Liberia; bans flights of Liberian-registered aircraft; bans the export of Liberian timber; broadens a still-current 1992 embargo on the export of arms and related material to Liberia; and bans foreign travel of senior Liberian government officials and their adult family members. The sanctions would remain in effect until the U.N. Secretary-General reports that the Liberian government is no longer assisting the RUF.[50] In addition to U.N.-focused efforts, and Administration inter-agency group is meeting

---

[49] Ambassador James B. Cunningham, Acting United States Representative to the United Nations, "Statement in the Security Council on the Panel of Experts Report on Sierra Leone Diamonds and Arms," January 25, 2001, *USUN PRESS RELEASE #11 (01)*.

[50] Judy Aita, U.S. Department of State, International Information Programs, "U.S. Pressing for Sanctions Against Liberia and Aid to RUF (Illicit diamond trade fuels Sierra Leone conflict, report says)," *Washington File*, January 25, 2001.

periodically, according to Department of State officials, to coordinate the development of U.S. policy on conflict diamonds. In testimony in October 2001 before the Committee on Ways and Means Subcommittee on Trade, Alan w. Eastham, Special Negotiator for Conflict Diamonds, U.S. Department of State and James E. Mendenhall, Deputy General Counsel, Office of the U.S. Trade Representative, outlined the Administration's conflict diamond policies. Both witnesses expressed Bush Administration commitment to working with the Congress on the conflict diamond trade, but both cautioned against legislation that might limit the Administration's ability to negotiate within the Kimberley Process.[51]

KIMBERLEY PROCESS

The Kimberley Process is an intergovernmental forum to which representatives of the diamond industry and non-governmental organizations contribute through a consultative process. The forum is developing and negotiating the design of an import/export certification system to govern international trade in rough diamonds in order to end trade in conflict diamonds. First sponsored by South Africa, it began as the Technical Forum on Diamonds, which met in May 2000 in Kimberley, South Africa. Four technical and ministerial meetings followed in 2000. At a meeting in Pretoria, South Africa in September 2000, the forum considered the interim findings of its Technical Working Group. It determined that a practical, reliable, and cost effective technical system for physically identifying the origin of individual diamonds did not exist, and recommended the establishment of an international export control regime, consisting of a system of sealed, registered diamond export parcels accompanied by forgery-proof certificates of origin, to be issued by exporting state authorities. The system would be overseen by a inter-governmental authority charged with monitoring and compliance, accreditation of national export regimes, and standard-setting, and possibly could be organized under U.N. auspices. It would also require the implementation of legal sanctions and penalties for violations of national-level legal export controls. The Working Group also noted a need for flexibility in any proposed system, especially vis-à-vis alluvial diamond mining and small-scale production and trading. It also recommended that participating nations ensure that domestic diamond marketing and production operate on the basis of open market competition

---

[51] Testimony available online [http://waysandmeans.house.gov/trade/107cong/tr-6wit.htm].

governed by a national system of transparency, disclosure and oversight of all diamond operations.

Conflict-Diamonds and the U.N. General Assembly

On December 12, 2000, the U.N. General Assembly (UNGA) adopted a resolution titled "The role of diamonds in fueling conflict: breaking the link between the illicit transaction of rough diamonds and armed conflict as a contribution to prevention and settlement of conflicts."[52] It was sponsored by 50 countries, including the United States. It called for measures to end the conflict diamond trade. The resolution recommended that a simple and workable international certification scheme for rough diamonds be created. Such a scheme, it stated, should be transparent, consistent with international law, and based "primarily on national certification schemes," that "meet internationally agreed minimum standards," and should not "impede...legitimate trade in diamonds or impose an undue burden on Governments or industry..." or compromise nations' sovereignty. Toward such ends, the resolution endorsed the Kimberley Process, requesting that it submit to the 56th session of the General Assembly a report on progress made.

Kimberley-Plus

In 2001, the process – dubbed the "expanded" Kimberley Process of "Kimberley-Plus" – continued, with meetings in Namibia, Belgium, Russia, the United Kingdom, Angola, and Botswana. The findings and formal recommendations of the Kimberley-Plus Process, represented by a draft document presently entitled the Kimberley Process Working Document, are slated to be presented in a report to the 56th Session of the U.N. General Assembly.

Overview of 2001 Kimberley Process Meetings

Windhoek, Namibia, February 2001

The first Kimberley-Plus meeting in Windhoek heard from Sierra Leone and Angola on the implementation of their respective diamond export certification systems; from Belgium on its import procedures, and from the HRD on proposed certification standards; and from Russia and Israel on their respective national diamond legislation. Based on these presentations

---

[52] A/RES/55/56, Ibid.

and a review of the initial technical and legal findings of the 2000 Kimberley meetings, a 'roadmap' defining the future focus and schedule of the Process was produced, and a Task Force was created to track progress and coordinate the work and meetings of the Process.[53]

Brussels, Belgium, April 2001

The Brussels meeting sought to identify elements of minimum acceptable standards for an international certification scheme, based on a review of existing import/export systems. It also examined the operational impact on business activities of a minimal standards certificate system, including the use of certificates in free trade zones and the feasibility of tracking diamonds after export from the original producing country.[54]

NGOs in attendance criticized the forum for making what they viewed as lack of substantive progress and for producing "bureaucratic wording and platitudes" due to what they saw as pressure from participants including the United States, Russia, Australia and the European Commission. They charged that "many government representatives stated that they had come to the meeting with no mandate to agree to anything, including even the most vague of wordings on issues that have now been discussed at five previous meetings."[55]

Moscow, Russia, July 2001

In early July, forum members reached preliminary agreement on the basic elements of a global certification system, including national responsibilities in confirming the legitimate origin of diamonds in countries of first export and re-export. Participants adopted a timetable, a working document, and a procedural framework to guide further Kimberley Process activities. During the meeting the World Diamond Council (WDC) proposed an industry-regulated chain of warranties system. It sought to standardize diamond industry trading practices and require its members to issue invoices guaranteeing the legitimate origin of their diamonds. The WDC suggested that such a system would prevent the growth of burdensome administrative processes, and state that it would "lobby governments to provide legal

---

[53] "Final Communiqué," Kimberley Process Meeting and Technical Workshop, Windhoek, Namibia, February 13-15, 2001.
[54] "Final Communiqué," Kimberley Process, Brussels, April 25-27, 2001.
[55] Fatal Transactions Campaign, "Kimberley Process Stalled?," April 27, 2001, [http://www.niza.nl/uk/press/FTpb010427.htm; and Other Facets [newsletter], "Kimberley Process Stalled? [*sic*]," June 2001.

endorsement for these warranties, with appropriate and enforceable penalties."[56]

Some non-governmental observers considered the meeting to be constructive, asserting that controversial issues were discussed openly, and that debate was not sidetracked by procedural roadblocks or excessive debate over nuances of draft language. Some viewed such progress as being motivated by proposed U.S. legislation. Some, however, noted a lack of progress on issues relating to national responsibilities for official import/export data collection and the maintenance of national certificate registries.

Twickenham, United Kingdom, September 2001[57]

In September, agreement in principle was reached on proposed key elements of an international certification scheme. It would require all exports of rough diamonds to be subject to a certification process backed by national rough diamond controls and procedures, a credible monitoring and oversight measures, and effective enforcement of certification scheme provisions. Under the proposed system, rough diamond shipments would be accompanied by forgery-resistant certificates and would be housed in tamper-proof containers. Participating countries would be required to collect and exchange rough diamond production and trade data, and to mutually exchange information about their respective national laws, procedures, and trade documentation.

The meeting also recognized the "useful role of self-regulation by the diamond industry which will fulfill minimum requirements." It also agreed to further assess the relationship between the proposed certification system and international trade law. European Union (EU) representatives at the meeting reportedly stated that the proposed system might conflict with EU common trade and customs policies such as provisions on open borders within the EU. Not all EU members are party to the Kimberley Process, and some observers questioned whether this would allow non-certified diamonds to enter the EU via non-Kimberley state and then pass freely into a Kimberley member country.

Non-governmental observers had mixed assessments of the meeting. They noted that many issues were covered in discussions, and viewed certain participating governments are exhibiting active commitment to rapid formulation of conflict diamond policymaking. However, they saw little

---

[56] WDC, Proposals for Self-Regulation with the Diamond Industry," July 2001.
[57] See "Final Communiqué," Kimberley Process Meeting in Twickenham, September 11-13, 2001.

progress reflected in formal revisions to the Kimberley Process Working Document, and asserted that provisions for implementing elements agreed to "in principle" were lacking. Ian Smillie, Research Coordinator for Partnership Africa Canada, stated that "there was lengthy debate about virtually every detail. In the end, there were actually more words, phrases and sentences left in brackets than before the meeting...a lot of what was agreed in principle was agreed in principle more than a year ago, and was debated all over again in London."[58] NGOs also rejected proposals for self-regulation by the industry and producing governments, contending that such measures alone would inadequately ensure the transparency of an international certification system. They called, instead, for an independent international certification monitoring mechanism.[59]

Luanda, Angola, October/November 2001

The Luanda meeting sought to finalize detailed proposals for an international rough diamond certification scheme. A Final Communiqué stated that "all alternatives and bracketed text had been removed from the working document, resulting in rapid progress being made." It stated that specific certification scheme agreements were reached on necessary national controls; required provisions to ensure co-operation and transparency; definitions of the precise certificate content; and on the necessity of consistency between the scheme and international trade law. Essential elements of a comprehensive system of warranties and industry self-regulation proposed by the World Diamond Council were also incorporated into the Working Document. These reportedly included external monitoring of its proposed chain of warranties. The EU delegation at the Luanda meeting reportedly issued a statement that all EU member states would be bound by the Kimberley agreement, once adopted. This reportedly ameliorated the fears of some observers about possible conflicts between common EU trade policies and those of individual member states.

Activists criticized the United States delegation to the Luanda meeting for deferring final agreement on key elements of the proposed system. The delegation reportedly entered reservations over proposals that diamonds should require certification during each international transshipment, suggesting instead, that certification might be necessary only during export from a producing country to a foreign buyer. The delegation also reportedly questioned whether certain proposals would comply with World Trade

---

[58] Other Facets, "Conflict Diamonds Are Forever? Major NGO Disappointment with Kimberley Process," October 2001.
[59] Other Facets, "Stop Blood Diamonds Now! The Key to Kimberley," October 2001.

Organization rules, and whether the creation of a secretariat to mange any eventual certification scheme would be necessary. Some observers also feared that other countries' Kimberley delegations would not allow finalization of measures to allow external monitoring of agreed internal controls.[60]

Gaborone, Botswana, November 27-29, 2001

The Gaborone meeting was the last formal meeting of the Kimberley Process in 2001; a further working meeting is planned, to be followed by an annual meeting in early 2002 after the submission of the Kimberley proposal to the UNGA. The Gaborone meeting endorsed a phased-in certificate of origin scheme. Nations prepared to undertake Kimberley-defined certified diamond trading would do so immediately; all others were encouraged to do so by June 2002, and an integrated international certificate system would become operational by the end of 2002. The meeting recommended extending the mandate of the Kimberley Process in order to finalize the proposed certification scheme, which it urged the U.N. to support.

The compatibility of the final recommendations of the Kimberley Process with World Trade Organization rules prohibiting restrictions on trade was reportedly a focal point of discussion. To ensure compatibility, some governments favored an open and non-exclusionary system allowing membership by all nations that assert that they meet certain minimal standards. Other participants advocated relatively strict membership standards, arguing that admission of states with weak regulatory schemes would undermine the system as a whole. They argued that some restrictions were permissible based on WTO rule exclusions that are said to allow for trade restrictions related to the protection of national security; others proposed seeking a waiver of certain WTO rules. A recommendation for a non-exclusionary system was adopted.

Several key participants cautioned that such a non-exclusionary system would require a robust monitoring and compliance system, which some

---

[60] UN Integrated Regional Information Networks, "Angola; Rough Diamond Certification In Progress," November 2, 2001; and Holly Burkhalter, "Blood On the Diamonds," Washington Post, November 6, 2001. A bracketed note in Section II of Kimberley Process working Document nr 8/2001, November 9, 2001, states that "The U.S. delegation [entered reservations] on, inter alia, Section II (a), Annex I and Section III (a) of the Working Document, which deal with the issuance of a Certificate to accompany every export of rough diamonds...[T]he delegation was not in a position to inform the other participants that, in case an agreement is reached, the U.S. would be able to issue a Certificate to accompany each export of rough diamonds from the United States."

reportedly believe did not emerge during the meeting.[61] Non-governmental observers asserted that at the Gaborone meeting "vitally important issues of compliance, verification and monitoring measures for new and current participants [were] considerably weakened...[and the meeting] failed to adequately address the crucial issue of a start date for the implementation of the scheme."[62] They criticized the meeting for neither establishing a secretariat nor setting out a budgetary framework for the funding of monitoring functions, making external monitoring of the system a function of consensus-based meetings of member states. They also criticized it for setting out limited requirements relating to the compilation and standardization by participating states of diamond trade and production statistics, and called for the implementation of the Kimberley Process proposals through a legally binding agreement sanctioned by a U.N. Security Council resolution.[63]

Prospects

Despite the recommendation in Gaborone that the certificate system be initiated, final agreement on several issues has yet to be reached, and some participants reportedly advocate further strengthening of certain measures related to monitoring of country compliance within the scheme.

CONGRESSIONAL ROLE

Congressional interest in achieving an end to the armed hostilities that have generated an international trade in conflict diamonds has motivated several legislative initiatives. These proposals have generally aimed at curtailing the ability of rebel groups fighting established governments to fund their armed activities through diamond export sales. Several hearings in both the House and Senate have addressed the issue of conflict diamonds in the context of hearings on U.S. policy on Sierra Leone, Angola, and U.N. activities in Africa. Among these hearings were:

---

[61] UN Integrated Regional Information Networks, "Deal Reached On "Conflict Diamonds," November 30, 2001; "DeBeers' Statement on Kimberley Process Agreements," November 29, 2001.
[62] Global Witness Partnership, Africa Canada and Fatal Transactions, "Kimberley Process Meeting: A Good Watchdog but Crucially Lacking Teeth," November 29, 2001.
[63] Ibid.; personal communications; and DailyNewsOnline, "NGOs Want Kimberley Agreement Amended," Republic of Botswana, 29 November 2001 [www.gov.bw].

- House Committee on International Relations, Subcommittee on Africa, *Africa's Diamonds: Precious, Perilous Too?*, May 9, 2000.
- House Ways and Means Committee, Trade Subcommittee, *Trade in African Diamonds*, September 13, 2000.
- House Committee on Ways and Means Subcommittee on Trade that was entitled *Conflict Diamonds*, October 10, 2001.

Each of these hearings addressed the human rights and conflict-related concerns motivating congressional interest in conflict diamonds. Witnesses also have highlighted commerce-related concerns, such as recommendations that proposed legislation be consistent with relevant World Trade Organization trade rules and with an emergent international diamond certification of origin trade regime. Witnesses have also called for legislation that does not penalize legitimate producers of diamonds, such as Botswana and South Africa. Some witnesses have pointed out that failure to enact legislation to curtail the conflict diamond trade and to introduce methods of separating legitimate diamonds from illicit diamonds might lead to a consumer-driven decrease in market demand for all diamonds, thus damaging the revenue base of legitimate diamond producing nations. In the October 2001 hearing, industry and non-governmental representatives described growing consensus between them on the need for a finalization of the Kimberley Process.

Legislation

As in the 106th Congress, several conflict diamond-related bills have been introduced in the 107th Congress. H.R. 918 (Hall) and S. 1084 (Durbin), both entitled the Clean Diamonds Act and H.R. 2722 (Houghton), entitled the Clean Diamond Trade Act, would prohibit diamond imports into the United States unless the exporting country is implementing rough diamond import and export controls that meet certain criteria. H.R. 918 and S. 1084 would prohibit the Overseas Private Investment Corporation (OPIC) and the Export-Import Bank (Ex-Im Bank) from engaging in certain transactions that assist countries violating the Act's requirements. On November 28, H.R. 2722, as amended, was passed by 408 to 6 under a motion to suspend the rules (Roll no. 453). On November 29, the measure was received in the Senate, read the first time, read the second time on November 30, and placed on the Senate Legislative Calendar under General Orders, Calendar No. 248.

S. 787 (Gregg), the Conflict Diamonds Act of 2001, would prohibit the importation, or abetment thereof, of diamonds exported from Sierra Leone, Angola, or Liberia, except for diamonds certified by the internationally recognized governments of Sierra Leone or Angola. It would prohibit imports into the United States of diamonds from countries that are not signatories to an international rough diamond import and export certification agreement, that are not implementing such an agreement, or that are not acting unilaterally to establish and enforce a similar certification system. S. 787 would authorize the President to prohibit imports of diamonds from Angola or Sierra Leone doing so is reasonably necessary to uphold conflict diamond-related UN Security Council resolutions. H.R. 918, S. 1084, H.R. 2722 and S. 787 urge the President to negotiate an international agreement to eliminate the trade in conflict diamonds.

Section 404 of S. 1215 (Hollings), the Senate version of the Appropriations bill FY2002, Commerce, Justice, State, would have enacted into law S. 787 of the 107th Congress (as introduced on April 26, 2001), but was indefinitely postponed in the Senate by unanimous consent. H.R. 2500 (Wolf), the Appropriations bill FY2002, Commerce, Justice, State offered by the House, would have included measures to prevent trade in conflict diamonds, but this language was "forced out of the bill on a procedural issue raised by the House Ways and Means Committee," according to Senator Gregg (*Congressional Record*, November 15, 2001, page S11881).

H.R. 2506 (Kolbe) would prohibit certain OPIC and Ex-Im Bank diamond-related projects in countries that are not implementing a system of rough diamond export and import controls, as defined in the Act. It would also prohibit the use of funds appropriated by the Act to assist countries that the Secretary of State determines, according to criteria outlined in the Act, to have actively destabilized the democratically elected government of Sierra Leone or aided or abetted illicit trade in Sierra Leonean diamonds.

APPENDIX: DIAMOND TRADE REGULATION IN COUNTRIES IN CONFLICT

Three African countries are the primary sources of conflict diamonds: Angola, the Democratic Republic of the Congo, and Sierra Leone. All have recently undertaken efforts to regulate the marketing and export of diamonds.

Angola[64]

Angola has been devastated by a lengthy civil war between the governing Popular Movement for the Liberation of Angola (MPLA), led by President Jose Eduardo Dos Dantos, and the rebel National Union for the Total Independence of Angola (UNITA), led by party president Jonas Savimbi. A series of broken peace accords and negotiations have caused the war to wane for limited periods, but armed hostilities have prevailed since Angola gained independence from Portugal in 1975. The most recent peace agreement, known as the Lusaka Protocols, fell apart during 1998, and by the end of that year full-scale war had resumed, and continues until the present.

In order to compel UNITA to comply with peach accords that it had signed during the Lusaka peace process, an existing U.N. arms and fuel embargo against UNITA was strengthened in June 1998 by the imposition of additional sanctions. U.N. Security Council Resolution 1173 required all U.N. member states to freeze UNITA assets; prevent the sale of Angolan diamonds lacking a government of national unity and reconciliation (GURN) certificate of origin; prevent the sale to UNITA of all vessels, aircraft and mining equipment. It also banned official contracts between U.N. member governments and UNITA, apart from those of the GURN, the U.N., Russia, Portugal, and the United States, which had acted as key mediators and observers in the Lusaka peace process.

Diamonds in Angola

Diamonds are found throughout Angola, but are notably abundant in the northeast of the country, in Lunda Norte and Lunda Sul provinces. Angolan diamonds are primarily gem-quality; the proportion of industrial grade diamonds has comprised between 10 and 15% of total diamond production in recent years, with the balance made up of near-gem and gem quality stones. Total production for 1999 is estimated to have been worth between $544 million and $619 million. The Diamond High Council reports that Belgium alone officially imported $58.24 million worth of Angolan

---

[64] For information on the conflict in Angola, see CRS Report 97-980, *Angola: Background and Current Situation*; CRS Report 98-816, *Angola Update*; and CRS Report RS20085, *Angola: War Resumes*. Periodic U.N. reports on the conflict are also online at [http://www.un.org/documents/repsc.htm].

diamonds in 1999.[65] A large proportion of Angolan diamond production – roughly two thirds – is artisanal or undertaken by small business firms.

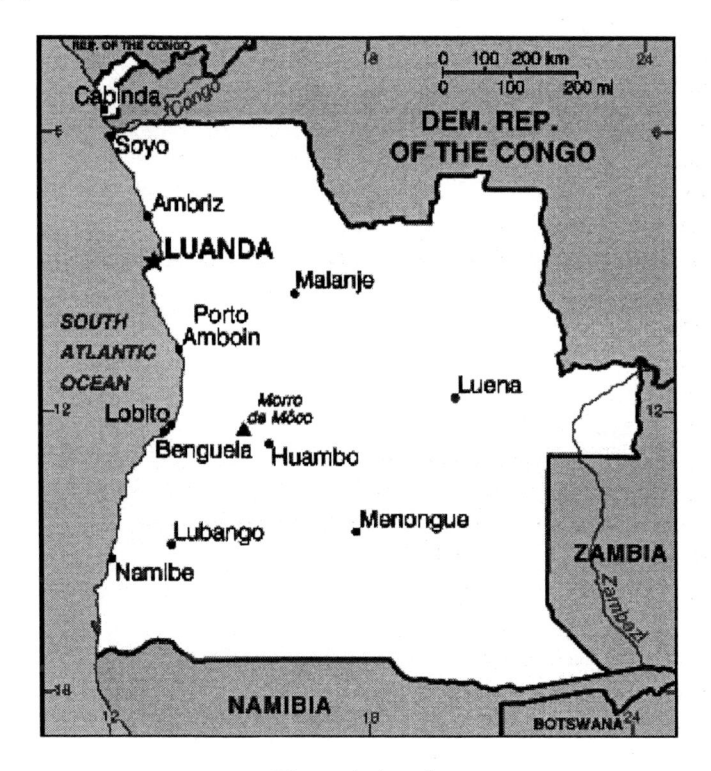

Figure 1. Angola

United Nations Sanctions Monitoring[66]

On March 10, 2000, a committee authorized by the U.N. Security Council to investigate the status of compliance with U.N. sanctions measures in effect against UNITA released a report of its findings. It documented violations of the sanctions by U.N. member states and private actors. The report focused on UNITA's marketing of diamonds in exchange of goods, services and logistical assistance in pursuing its war against the Angolan government. It also showed how UNITA had used diamonds to maintain a

---

[65] HRD Diamond Council, Quarterly Trade Statistics, available online at [http://www.diamonds.be/main11.htm].

[66] *Report of the Panel of Experts on Violation of Security Council Sanctions Against UNITA, op. cit.* The report has been dubbed the *Fowler Report* after the name of the sanctions committee chairman, Canadian U.N. Ambassador Robert R. Fowler.

network of international political support. The report detailed actions in support of UNITA by the governments of Presidents Gnassingbé Eyadéma of Togo, President Blaise Compaoré of Burkina Faso, and former President Mobuto Sese Seko of the former Zaire (now the Democratic Republic of Congo, or DRC). Such actions included permitting and substantively aiding in the transfer of arms through each country, facilitating meetings between arms and diamond dealers and UNITA, and stockpiling weapons on behalf of UNITA.

The chapter also described links between UNITA and government officials in Rwanda, Zambia, Côte d'Ivoire, Congo-Brazzaville, and Gabon that had facilitated UNITA's diamond and weapons deals. The report asserted that France, Portugal, Switzerland, Namibia, South Africa, the United States, and several other countries had allowed UNITA representatives greater freedom to operate or travel in each country than is permitted by the sanctions regime. The report named Bulgaria as a principal source of UNITA arm shipments, Ukraine as a possible source, and criticized Belgium for laxity in controlling the origin of imported diamonds. The report made nearly 40 detailed recommendations suggesting ways of curtailing UNITA's diamond trading, punishing sanctions transgressors, and regulating the diamond trade. A five-member committee panel is slated to follow up on the report's findings.

In July 2000, acting in response to U.N. Security Council Resolution 1295 (2000), U.N. Secretary General Kofi Annan established a Monitoring Mechanism to monitor compliance with U.N. resolutions sanctions against UNITA. On October 25, 2000, the Chairman of the U.N. Security Council's Angola Sanctions Committee distributed to the Chairman of the Security Council an interim report (S/2000/1026) that described monitoring of compliance with U.N. sanctions against UNITA by the Sanctions Committee's Panel of Experts. The interim report used the *Fowler Report* as a point of departure. The interim report described investigatory progress by the Panel of Experts to date, and summarized its preliminary findings. In some instances, e.g., in relation to arms trading, the Panel found continuing "serious discrepancies" in the information received from the arms importing and exporting countries. It also found mixed levels of cooperation by member states in prohibiting UNITA travel and representation. In several cases, after being formally banned, UNITA representatives had formed "front" organizations that member states "found it difficult to prohibit." Despite significant international action aimed at banning the trade in conflict diamonds, the Panel found that UNITA continues to be able to mine and sell diamonds. It also described significant action taken by the government of

Angola aimed at establishing a diamond export certification system. In late December 2000, the Monitoring Mechanism published a final report on its investigations and findings. The report, U.N. S/2000/1225, augmented the Mechanism's interim report, adding considerable detail, and contained a range of recommendations aimed at tightening the sanctions against UNITA. On January 23, 2001, the U.N. Security Council, taking note of the report, extended by three months the mandate of the Angola panel of experts.[67]

Regulation of the Angolan Diamond Industry

In order to comply with U.N. resolution 1173, the Angolan government is currently implementing a system to prevent conflict diamonds from entering the legal market by better regulating the mining and sale of diamonds in the country, thereby denying UNITA a domestic market. Beginning in early 2000, Angola instituted a diamond marketing and certificate of origin system designed to allow the government to guarantee that conflict diamonds are not part of officially-sanctioned exports; cut off sources of funding for UNITA; and to increase government tax revenues. Under the system, the majority of Angolan diamonds are sold through a single company, the Angolan Selling Corporation (Ascorp).

Ascorp

Ascorp is a joint venture between Sodiam UEE, a company owned by the Angolan government and two private foreign investors. Sodiam was established by the Angolan government's Ministry of Geology and Mines and Endiama, a government agency charged with licensing and regulating diamond mining which, until the creation of Ascorp, held the sole right to market Angolan rough diamonds. Sodiam reportedly owns 51% of Ascorp. The remainder, as reported by Africa Energy and Mining and the Financial Times, is owned roughly evenly by Lev Leviev, an Israeli diamond buyer with interests in the Russian diamond industry and by Belgian business interests, which include the Omega Diamond firm. Most other buying offices in Angola have been ordered to shut down. Ascorp's foreign joint partners were reportedly chosen because they possess substantial operating capital, extensive international trade connections, and trade independently of De Beers. A small Australian company called Majestic Resources, however, announced in April 2000 that it had signed a contract to sell $20 million worth of official Angolan diamond sales.

---

[67] United Nations Security Council, "Resolution 1336 (2001) Adopted by the Security Council at its 4263rd meeting, on 23 January 2001, January 23, 2001, [http://www.

Domestic Regulatory Reform

In order to prevent conflict diamonds from entering official state channels, Angola is reportedly registering between 300,000 and 350,000 artisanal miners and small-scale diamond traders. This effort is aimed at establishing a system of national-level documents that will enable the government to maintain a chain of legal provenance for diamonds produced by small-scale miners and purchased by official, government-licensed diamond buyers. The scheme aims at cracking down on second-tier, informal market buyers from whom some official buyers have reportedly purchased diamonds. Observers believe that these intermediary buyers form a conduit that in the past has allowed diamonds from rebel-controlled areas to enter the official diamond market. To enforce the Ascorp system, a special police and judiciary system with law enforcement powers is being created to regulate the artisanal diamond industry.

The government has reportedly dramatically increased its revenues following implementation of the Ascorp scheme, which has drawn complaints from large diamond brokering and sales companies, such as De Beers and other firms formerly held Angolan diamond mining and selling rights. In May 2000, De Beers and some smaller companies attempted to freeze a shipment of Angolan diamonds to Antwerp, which they claimed was theirs under prior contract agreements. De Beers has reportedly claimed that it holds a contract, backed by a loan worth about $50 million, with Sociedade de Desenvolvimento Mineiro (SDM), a joint venture between Endiama and two foreign mining and construction firms, to purchase the equivalent of 20% of Angolan rough diamond official sales.[68]

Anecdotal accounts allege that Ascorp is using its monopsony powers to pay sub-market prices to artisanal miners, and to have those who refuse arrested, as it is illegal to sell diamonds to buyers other than Ascorp. Ascorp officials claim that rising government diamond revenues and the decrease in diamond-rich territory held by UNITA, as a result of recent battles and earlier government battlefield victories, indicate that exports of Angolan conflict diamonds are decreasing. However, on-going conflict makes effective administration of large regions of Angola impossible, and the alleged payment of sub-market prices to producers may motivate smuggling. In addition, UNITA representatives have recently reportedly stated in early October 2000 that it controls half of Angola's diamond producing areas. As

---

un.org/Docs/scres/2001/res1336e.pdf].
[68] "Diamonds: The Angolan question," *The Mining Journal*, March 31, 2000, p. 243.

a result of these factors, the possibility of UNITA-mined diamonds entering the official diamond market continues to exist, industry watchers report.

Democratic Republic of Congo[69]

In August 1998, a rebellion began in the Democratic Republic of Congo (DRC) against the government of Laurent Kabila. Kabila had come to power as the head of an armed alliance backed by Uganda and Rwanda, the Alliance of Democratic Forces for the Liberian of Congo-Zaire (ADFL). The ADFL seized power in May 1997 from Mobutu Sese Seko, who had ruled the country since 1965. The war continues until the present. It is complex. At least three rebel movements backed by Uganda, Rwanda and to a lesser extent by Burundi are fighting the DRC government and its allies. The allies currently control between one-third and one-half of the country. The DRC government's allies presently include Zimbabwe, Angola, and Namibia. Chad and Sudan have also reportedly fought, or provided logistical support, on behalf of the DRC government. In addition, several militias in eastern DRC are allied with the government.

Diamonds in the DRC

Diamonds are principally found in two regions of the DRC: East and West Kasai Provinces, especially around the towns of Tshikapa and Mbuji-Mayi, and to a limited extent in the region surrounding the city of Kisangani, in Province Oriental and in Bandundu Province. Between 70 and 90% of Congolese diamonds are industrial-quality; the remainder are jewel-quality gems. More than half of diamond mining is artisanal. Total production for 1999 is estimated to have been worth between $396 million and $759 dollars million or more. The Diamond High Council reports $758.75 million worth of diamonds were imported to Belgium from the DRC in 1999.

Diamonds and Conflict in the DRC

Analysts do not regard the Congo conflict as primarily a war over diamonds. Control of diamond areas, however, has played an important role in strategic terms, since control over these areas simultaneously yields the possibility of a significant source of funding and denies access to the same resources by opposing forces. Diamonds may also represent a source of

---

[69] For additional background on the conflict in the DRC, see CRS Issue Brief IB96037, *Congo (Formerly Zaire)*. Periodic U.N. reports on the conflict are also available at [http://www.un.org/documents/repsc.htm].

frictions among allies. Several battles between Rwandan and Ugandan forces in Kisangani are reportedly attributable, in part, to conflict over control of diamond and gold mining in the region.[70]

Figure 2. Democratic Republic of Congo

Diamonds may be a significant factor enabling the Kabila regime to maintain foreign military support. One expert has reportedly estimated that during the presidency of the late President Laurent Kabila, DRC government diamond-based earnings totaled approximately $800 million.[71] Many press accounts have reported that Zimbabwean military assistance for the Kabila government is contingent, in part, upon the receipt of exclusive and valuable business concessions in the DRC by Zimbabwean firms. Some of these firms are reported to have close ties to the Zimbabwean military and ruling party,

---

[70] See, for instance, Lara Santoro, "Behind the Congo war: diamonds," *Christian Science Monitor*, August 16, 1999, [http://www.csmonitor.com/durable/1999/08/16/pls4.htm] and Gunnar Willum, "Rebel Leader Confirms What Western Donors Deny: Uganda plunders Congo, *Aktuelt* [Denmark], January 22, 2001.

[71] *Reuters*, "Blood Diamonds Fueled Fighting in Kabila's Congo," January 17, 2001.

and diamond mining rights have played a high-profile role in these alleged business arrangements. In October 1999, Cosleg, a joint venture between Osleg (acronym for Operation Sovereign Legitimacy), a company owned by the Zimbabwean Defense Forces, and Comiex, a company owned by the DRC army with links to the Presidency, was announced. The companies reportedly planned to purchase artisanal gold and diamond production, and possibly in the future undertake mining, timber, and other extractive operations directly.

Cosleg reportedly prepared to initiate mining operations at a substantial mining concession west of Mbuji-Mayi in south-central DRC that was previously owned by Societe Miniere de Bakwanga (MIBA). MIBA is a parastatal that is the country's largest diamond producer and the former official holder of monopoly diamond mining rights in the DRC. The Cosleg operation reportedly involved the use of Zimbabwean earth moving equipment under the control of the Zimbabwean Defense Forces. It was supported by technical and financial assistance provided by Oryx Zimcon, a company reportedly financed by the Omani Consul in Zimbabwe, Kamel Khalfan. Khalfan reportedly runs a range of businesses in Zimbabwe and has alleged close ties to ZANU-PF. Oryx Zimcon and Osleg reportedly are both registered with addresses at the Zimbabwe Ministry of Defense of Harare.[72]

Oryx Listing

The two Congolese and Zimbabwean military companies and Oryx Zimcon later entered into a business agreement with an Omani-linked firm called Oryx Natural Resources. Plans called for the latter to be listed on the London Alternative Investment Market (AIM) as Oryx Diamonds, through a reverse takeover of a listed company registered in the Cayman Islands, Petra Diamonds. The move would have allowed Oryx to raise funds or acquire other firms through its listing, in part based on Oryx's reported rights to a substantial mining concession near Mbuji-Mayi. In June 2000, former American Ambassador to Burundi Frances Cook was reported to be a proposed member of the board of directors of the newly listed company.[73]

---

[72] "Glittering prizes from the war," *Africa Confidential*, May 26, 2000, Vol. 41: 11, Chris Gordon, "Zimbabwe army seeks payback in Congo," *Mail and Guardian* online edition, Oct. 1, 1999, online at [http://www.mg.co.za/mg/news/99oct1/1oct-congo.html]; and Lawrence Bartlett, "War and riches inspire new African economic deal making," *Mail and Guardian* online edition, Sept. 2, 1999, [http://www.mg.co.za./mg/news/99sep2/30sep-congo.html/, among other reports on the Zimbabwean role in the war in the DRC.

[73] "Change at the top at MIBA," *Africa Mining and Energy*, No. 277, June 4, 2000; and "Petra Diamonds Ld – Proposed Acquisition, etc.," *Regulatory News Service/World Reporter*, May 18, 2000.

Profits were reportedly to be split between Petra Diamond shareholders and the Cosleg partners. The planned listing was later withdrawn, reportedly under pressure from NGOs, the London Stock Exchange, and Britain's Foreign Office. They objected to the deal on the basis that the concession, worth a reported $1 billion, was located in a conflict zone and was being undertaken by parties to the DRC conflict. The listing may, reportedly, be undertaken in the future, possibly in Ireland or North America.

Namibian Government Mine in the DRC Reported

In late February 2001, the Namibian and international press reported that a Namibia Ministry of Defense parastatal corporation, August 26 (Party) ltd., had been allocated a 25-square kilometer diamond concession at Maji Munene, DRC, close to the Angolan border near the DRC diamond mining town of Tshikapa. The proposed mining operation will reportedly be run by a joint business partnership involving an unnamed U.S. firm and the Congolese and Namibian governments. Namibian opposition parties have harshly criticized the government for using its military presence in the DRC to "plunder" Congolese diamonds and assert that diamonds derived from the deal comprise conflict diamonds. They have also criticized the Namibian government for not revealing the facts or existence of the reported deal, including in response to direct Parliamentary inquiries. The membership of the Board of Directors of August 26 Company reportedly includes top Namibian military and government officials, and press reports have tied the mine to Namibian President Nujoma. The government has defended the mine concession as a "bilateral trade and investment agreement."[74]

Regulation of the Congolese Diamond Industry

In response to publicity surrounding the Oryx deal and increased international public concern about conflict diamonds, the DRC government has over the past year acted to more closely regulate the diamond industry. Apart from its concern about conflict diamonds, the DRC government has also expressed a desire to gain increased control over diamond trading and mining, crack down on smuggling, and increase diamond-related revenue collections.

---

[74] See Associated Press, "Outrage After Namibian Government Owns Up to Diamond Interests in the Congo," February 24, 2001; Agence France Presse, "Namibia Admits Involvement in Diamond Mine in DR Congo," February 22, 2001; Tangeni Ampuadhi, "Government Confesses About Congo Kinshasa Gem Mine," The Namibian, February 23, 2001; and Panafrican News, "Namibia's Diamond Deal in DRC Exposed," February 22, 2001; and Deutsche Presse-Agentur, "Namibia has Mine in the DRC Admits Mines Minister," February 22, 2001.

The government's actions also respond to U.N. Security Council concerns about the link between conflict and mineral resources in the DRC. The Council in early June 2000 requested the U.N. Secretary General to set up a panel of experts on the illegal exploitation of natural resources and other forms of wealth in the DRC. On August 17, 2000 the DRC government reportedly filed with that panel a report alleging that Twandan and Ugandan companies are illegally expropriating and expatriating Congolese gold, diamonds, and other precious minerals. The report was reportedly compiled by Observation Gouverance Transparence in Kinshasa, an NGO, in April 2000.

In mid-January 2001, the U.N. panel of experts submitted to U.N. Secretary-General Kofi Annan an interim report of its findings. The panel – which met or attempted to meet with parties to the DRC conflict, including governments and rebel officials – reported having received "varying levels of cooperation from its interlocutors, ranging from apparent openness to near hostility."[75] The panel described it investigatory challenge:

> One of the most serious problems facing the Panel is the paucity of detailed and reliable information, including statistics, as to the nature, extent, location, yield and value of the natural resources of the Democratic Republic of the Congo. Decades of Government neglect, mismanagement and corruption, including widespread evasion of taxes and customs duties, not to mention the effects of conflict since 1996, make it almost impossible to establish a precise and impartial factual picture of the country's natural resource base and exploitation patterns. Though rumor and anecdote abound, documentary evidence is almost non-existent. Mines and other sources of natural wealth are remote and heavily guarded, often located in areas subject to outbreaks of fighting or armed attacks against the local population. Roads are few and ill maintained and communications poor. The Panel has found in its own investigations that activity around the mines is cloaked by an atmosphere of lawlessness, violence and fear.[76]

The panel met with government officials from the DRC, Kenya, Uganda, Rwanda, Burundi, and Zimbabwe, foreign diplomats based in the above countries, non-government officials, civilians, and U.N. organization officials in the region. The panel was told of declining extraction and production of natural resources in area of the DRC held by the country's

---

[75] United Nations, "Annex: Interim Report of the United Nations Expert Panel on the Illegal Exploitation of Natural Resources and Other Forms of Wealth in the Democratic Republic of the Congo," S/2001/49, page 3.

[76] Ibid., page 4.

government, which accused Rwanda, Uganda, and its rebel allies of plundering the country's natural resources, including gold, diamonds, Colombo-tantalite, and agricultural products, appropriating power generation plants and capacity, and of massacring protected species. Uganda and Rwanda denied such claims and sought to explain the national security reasons for their military interventions in the DRC. The panel of experts has reportedly been criticized by members of the U.N. Security Council for failing to include a focus on "commercial and research groups" during its investigation.[77] The Security Council is also reported to have refused to extend the March 2001 deadline for submission of the panel's final report.

IDI Diamond Monopsony[78]

In the latter half of the year 2000, the DRC government undertook an initiative aimed at controlling and regulating its diamond trade similar to that taken by Angola. On July 31, 2000, according to press accounts, the DRC government signed an 18-month monopoly contract with IDI Diamonds, an Israeli rough diamond-trading firm owned by businessman Dan Gertler. The contract gives IDI exclusive rights to buy and market uncut Congolese diamonds, both from MIBA and from all other authorized trading firms. The contract, which is limited to territory controlled by the DRC government, requires IDI to pay fees of $20 million annually to the government for exclusive control over trade wroth an estimated $600 to 700 million dollars annually. The IDI-DRC contract creates a quasi-public corporation, Societe de Development du Diamante (SDD), also reportedly called International Industries Congo, owned jointly by IDI (30%) and the DRC government (70%). The contract revoked all existing export and buying licenses within thirty days of its signing, although existing trading companies will reportedly be able to operate under unspecified standards to be established by SDD.

News reports state that the object of the arrangement with IDI, according to the government, is to maximize revenue collections, which have fallen recently due to increased smuggling in the wake of a government order in 1999, which banned foreign participation in diamond buying and required market transactions to be undertaken in Congolese francs. It is also meant to provide the means to establish the legal provenance of diamonds

---

[77] The Mining Journal, "WDC maintains momentum; Congo report lacks data," January 26, 2001, page 65.

[78] This account of the IDI diamond deal draws upon coverage of multiple issues of *Africa Energy and Mining, the Financial Times* of London (Online at [http://news.ft.com]); and articles from Tacy Ltd., a diamond industry consulting firm, online at [http://diamond consult.com/info.htm], *inter alia.*

exported from Congo in order to certify that they are from areas held by the internationally recognized government of Congo. In addition, it is aimed at improving the regulatory capacity of the Centre National d'Expertise (CNE), a government organ charged with setting up a diamond certification system and evaluating diamond values and weights. As part of the IDI deal, the government stated that it would issue certificates of origin for all rough diamond exports to prevent conflict diamonds from entering official trade. IDI will also reportedly establish representation in the diamond areas held by the government, including in the Tshikapa area and the nearby border area with Angola in order to monitor and verify the legitimate origin of diamonds entering its stock.[79]

Past Regulatory Efforts

The IDI deal comes in the wake of earlier DRC government attempts to rationalize and control the country's diamond industry. In mid-2000, CNE had reportedly consulted with Zurel Brothers, an Antwerp firm, to assist the DRC government in setting up a diamond certification and production control system. The status of the results of this consultancy in the contest of the IDI deal is not clear.

In May 2000, the government reportedly opened a training center with the objective of creating a special mining police force that will be able to halt theft at mines, regulate artisanal mining and informal diamond traders, and prevent the under-evaluation of diamonds by CNE officials. In early April 2000, the DRC government announced plans to implement a new mining code to attract western investors to the country. In 1999, the government issued regulations that centralized all diamond trading in Kinshasa; banned the presence of foreigners in diamond mining regions and the purchase of diamonds with foreign currency; and required foreign traders to pay license fees of between $100,000 and $150,000 per license. Some industry observers believe that these measures have contributed to sharp drops in the volume and dollar value of diamond exports in 2000. The local affiliate of one firm, Lazare Kaplan International, was given an exemption in April 222 that

---

[79] Press accounts reported that the deal involves the promise of training of DRC armed forces by the Israeli army, in order to prevent cross-border smuggling of diamonds. This was denied by the Israeli Defense Ministry and by Israeli Diamond Exchange president Schmuel Schnitzer. Schnitzer stated that IDI might provide recommendations relating to security firms to assist the DRC armed forces to combat smuggling but will not be directly involved in anti-smuggling training or operations. See "Israel defends controversial diamond deal," *Agence France-Press*, Sept. 7, 2000; "Congo denies Israeli trainers offered in exchange for diamond deal," *The Jerusalem Post*, Sept. 6, 2000, p. 12, and "Congo

allowed it to set up trading operations in the Tshikapa region and to trade in foreign currency, an action that reportedly infuriated local diamond dealers.

Prospects

There are indications that the IDI deal may not endure. In late August 2000, nine DRC diamond-trading houses signed a joint letter to the government requesting that their licenses-revoked earlier in the month - be reinstated, and that they continue to be permitted to export diamonds. According to news reports, diamond exports were halted following the revocation of local traders' permits. The lack of sales was reportedly due to a refusal by traders to sell their diamonds to IDI and to a "slow start" by IDI in making diamond purchases. The charge that IDI was not implementing its monopoly rights rapidly and that it was the primary cause of the currency shortage has been denied by Shmuel Schnitzer president of the Israeli Diamond Exchange. The Panafrican new agency reported on September 19, 2000 that a DRC ministerial meeting had upheld the IDI contract, and that the ministers had requested IDI to speedily set up diamond purchasing centers throughout the country.

To address the complaints of diamond traders, the DRC government was reported to be setting up an arbitration agency, the Service for the Development of the Congolese Diamond, to mediate price disputes between IDI and domestic diamond traders. The lack of diamond exports following the IDI deal reportedly caused a drastic foreign currency shortage and a steady and sharp drop in the value of the Congolese franc on the black market, from a rate of around 65 francs per U.S. dollar to 95 francs as of mid September 2000. The local currency, pegged at an official rate of exchange of 23.5 Congo francs per dollar, has been devalued three times since early 1999, when commercial transactions for foreign currency were banned in order to stabilize the local currency and to maintain low fuel and other basic commodity prices. The DRC has reportedly lacked foreign currency reserves in recent months, and the shortage has caused a contraction of imports in al economic sectors, reportedly causing a sharp contraction in economic activity and shortages of fuel and staple food.

Recent Developments

The DRC government is reportedly attempting to counter recent negative economic trends, and increase economic transparency, market

---

signs Israeli company to stop smuggling: Firm denies military training," *National Post* (Canada), Sept. 5, 2000, p. A10.

competition, and foreign investment in the Congolese economy. In early February 2001, the new DRC government of Joseph Kabila (who took power following the assassination of his father, Laurent Kabila) was reportedly preparing to lift the ban on free circulation of foreign money and liberalize the diamond market and exchange rate regime. A new mining code, scheduled for release in April 2001, is reportedly being developed with the assistance of the World Bank. The new mining code will reportedly be based on a uniform, first-come, first-served licensing system in place of case-by-case negotiations and will not required presidential authorization. Instead, mining titles will be granted by the ministry of mines, and commercial registration procedures will be simplified. DRC officials state that the reforms are aimed at transforming an unpredictable business climate in which mining rights have been granted and taken away in an apparently arbitrary fashion in order to increase foreign investment. A new tax system based on "reasonable rates" and consideration of profitability will also reportedly be introduced. The proposed mining code is also said to include environmental measures aimed at controlling mining waste treatment and will be accompanied by a small-scale mining development program.[80]

Sierra Leone[81]

From 1991 until the present, successive governments of Sierra Leone, a small West African country with significant mineral and timber resources but a poorly developed economy, have been besieged by a rebel group, the Revolution United Front (RUF). Two inter-related factors motivate the hostilities:

- The growth of systemic government corruption leading to a severe deterioration in the capacity of the state to govern and provide basic public goods and services. In this sense, the conflict has domestic roots.

---

[80] *The Mining Journal*, "DRC lifts foreign cash ban," February 2, 2001, page 88; RTNC TV, Kinshasa, "DRCongo: President Joseph Kabila's speech on day of inauguration," *BBC Monitoring Africa – Political*, January 27, 2001; and Nicol Degli Innocenti, "Congo Hopes Mind Code Will Boost Investment," *Financial Times*, February 9, 2001, page 36.

[81] The Sierra Leone conflict is described in CRS Report RL30367, Sierra Leone: Background and Issues for Congress; CRS Report RS20578, Sierra Leone, A Failed Peace? And in CRS Issue Brief, IB96025, Liberia: Current Issues and United States Policy. Periodic U.N. reports on the conflict are also available at [http://ww.un.org/documents/repsc.htm].

- Conflict over control of mineral wealth – particularly diamonds – and state resources, with participation by domestic and foreign actors. In this sense, the conflict can be seen as a result of both domestic factors and of long-term political instability in the Mano River sub-region of West Africa.

The RUF is infamous for its brutal treatment of civilians. In the first half of 2000, following the UN's assumption of peacekeeping duties, the RUF kidnapped over 500 U.N. peacekeepers and military observers. All were released but the U.N. operation continues to receive sharp criticism for its failure to implement its peacekeeping mandate. Internal discord has also been reported within UNAMSIL, the U.N. peacekeeping mission in Sierra Leone.[82]

The configuration of the sides in the Sierra Leone conflict has shifted periodically. Relations between individual units of the SLA and the RUF have ranged from mutual non-interference to outright collaboration. This has led to the coining of the term "sobel," i.e., "soldier-rebel." Access to diamond mining areas in the country's southeast and the looting of civilian property and other resources have often been the apparent purpose of such cooperation.

Liberian Role

The origin and operational capacity of the RUF have been closely tied to the politics of Liberia. In the first years of the Liberian civil war, the National Patriotic Front of Liberia (NPFL), headed by Liberia's current President, Charles Taylor, reportedly sought to assert control over the trade in diamonds and other resources in the area near the Liberian-Sierra Leone border and to prevent rival groups from exploiting this trade. The NPFL also is reported to have recruited fighters in Sierra Leone and reportedly assisted the Revolutionary United Front (RUF) in its fight against successive Sierra Leonean governments. The historical tie between Mr. Taylor and the RUF is widely believed to have endured. Liberia continued to provide assistance, such as arms and logistical support, to the RUF in exchange for smuggled diamonds and other resources, according to many news reports and statements by concerned governments. These actions, for which President

---

[82] Douglas Farah, "Old Problems Hamper U.N. in Sierra Leone; Leadership, Equipment Troubles Leave Peacekeepers Vulnerable," *Washington Post*, June 11, 2000, p. A25; and Douglas Farah, "Internal Disputes Mar U.N. Mission," *Washington Post*, September 10, 2000, pp. A1, A34.

Diamonds and Conflict: Policy Proposals and Background 47

Taylor and his ministers have repeatedly denied responsibility, are said to have considerably hampered efforts to end the conflict in Sierra Leone.

Figure 3. Sierra Leone

In March and June 2000, the international press reported the existence of fresh intelligence held by western and regional governments that confirmed the RUF-Liberian ties. Airplanes ferrying wounded fighters, field supplies and food, diamonds, and arms were reported to be flying regularly between core RUF areas in western Sierra Leone and Monrovia, the Liberian capital, and other areas of Liberia. Radio traffic between RUF units, referring to Liberian logistical support, and between the RUF and Liberian officials, was also reportedly intercepted. Arms for the RUF are also reported to have been imported into Burkina Faso, and then trans-shipped to Liberia for delivery to Sierra Leone in exchange for diamonds. American officials have repeatedly warned Taylor directly that they believe that he is supplying the RUF with arms in exchange for diamonds, and have explicitly called on him to end the relationship. Likewise, Britain, which has sent troops and military trainers to assist the government of Sierra Leone (GOSL) against the RUF, has taken an especially direct stance on the issue of alleged Liberian assistance to the RUF in exchange for diamonds. It succeeded, in mid June 2000, in suspending approximately $50 million of European Union (EU) aid to Liberia, against considerable opposition of other EU members. The Liberian government has repeatedly and consistently denied that it is a conduit for RUF diamonds.

Diamonds in Sierra Leone

Diamonds are found in approximately one-third of Sierra Leone's territory, principally in the east and southeast. Alluvial deposits (those found on or near the surface and in river beds) predominate, but several diamond-rich kimberlites (deep pipes, of volcanic origin, made up of eruptive tock materials) are also present. Artisanal mining, which ahs been an important mode of extraction since diamonds were discovered in Sierra Leone, is currently the primary source of production in the country. Conflict has largely halted production by large scale, formal sector mining firms, which have historically contributed significantly to production.

Sierra Leone Diamond Exports

Diamonds have historically been a key source of foreign exchange for Sierra Leone and have accounted for between 80 and 90% of export earnings in recent years, although some of this hard currency has reportedly entered the country via the black market. Liberia and Guinea have been named as primary destinations for smuggled Sierra Leone diamonds, as has Burkina Faso to a lesser extent.[83] Diamond production, predominantly unofficial or illicit, has been estimated as being worth between $70 and $138 million dollars in 1999, but annual production may be significantly higher. Some observers have attributed a major source of a rapid rise in Liberian diamond exports, totaling more than $298 million in 1999 and $268 million in 1998, to diamonds smuggled from Sierra Leone to Liberia. Others claim that the rise is attributable to the transshipment and re-export of diamonds from Russia and elsewhere through Liberia in order to avoid Belgian import tax payments. Liberia has a reported annual diamond production capacity of 100,000 to 150,000 carats.

Official Sierra Leone diamond exports in 1999, according to the Sierra Leone Government Gold and Diamond Office (GGDO), as reported by

USAID, were worth a total of $1.5 million dollars.[84] The Diamond High Council, of Antwerp, Belgium, however, has reportedly stated that uncut diamonds officially exported from Sierra Leone were worth $31 million in 1999, down from $66 million in 1998.[85] From January through June 2000, the GGDO reported that diamonds valued at $3.45 million were exported from Sierra Leone.[86] No diamonds were legally exported from Sierra Leone following the passage by the U.N. Security Council of Resolution 1306 (2000), which prohibits the exports of Sierra Leonean diamonds lacking GOSL certification under a UN-accredited certification scheme, according to USAID.[87]

Regulation of Sierra Leone Diamonds

The United Nations, through Security Council Resolution 1306 (2000), has prohibited for 18 months the importation of U.N. member states of all rough diamonds from Sierra Leone not covered by a certificate of origin regime implemented by the GOSL.[88] The resolution calls on the government to set up such a system with help from U.N. member states and international organizations. All diamonds certified by the GOSL, following official creation of a functioning regime, are exempt from the prohibition. The resolution also calls upon the Sanctions Committee on Sierra Leone created by Resolution 1132 of 1997 to take measures to identify persons or entities engaged in violations of the diamond export sanctions and to report to the Security Council on such findings. On January 18, 2001, then-President Clinton issued Executive Order 13194, "Prohibiting the Importation of Rough Diamonds From Sierra Leone." The intended effect of the action was to implement U.N. Resolution 1306, prevent the RUF from profiting from the sale of diamonds in the United States, and support the legitimate trade in diamonds.

---

[83] Interview with Sierra Leone Ambassador to the United States, John Leigh, September 18, 2000; and Tim Sullivan, "Rebel diamonds turn to guns in West African smuggling network," Associated Press Newswires, August 5, 2000, *inter alia.*

[84] USAID, *Diamonds and Armed Conflict in Sierra Leone: Proposal for Implementation of a new diamond policy and operations*, Office of Transition Initiatives, May 8, 2000, online at : [http://www.usaid.gov/hum_response/oti/country/sleone/diamonds.html].

[85] Personal communication with HRD Diamond Council, *inter alia.*

[86] *"Diamond export statistics,"* Presentation by the Government of the Republic of Sierra Leone to the United Nations Sanctions Committee, July 31, 2000.

[87] Interview with USAID official, Sept. 22, 2000.

[88] United Nations Security Council, "Resolution 1306 (2000) Adopted by the Security Council at its 4168th meeting, on 5 July 2000," S/RES/1306 (2000), July 5, 2000, [http://www.un.org/Docs/scres/2000/res1306e.pdf].

On July 31 and August 1, 2000, the Panel of Experts called for in Resolution 1306, made up of members nominated by the Sanctions Committee, held exploratory hearings on the illicit diamond trade. The hearings focused on the alleged role of Liberia and Burkina Faso in the conflict diamond trade out of Sierra Leone, the reported role of Ukraine in supplying arms to the RUF via Liberia and Burkina Faso, and the GOSL's proposed diamond export certification plan. At the hearing the Sierra Leone plan was criticized by Monie Captan, foreign minister of Liberia who, according to the *Financial Times*, "questioned which diamonds the GOSL would be certifying since it did not control the mines." Representatives from the United Kingdom, the United States, and NGOs reportedly voiced concern that, in spite of the proposed certification scheme, conflict stones might enter into legitimate trade. Following the hearings, the Sanctions Committee's Panel of Experts undertook field investigations into illicit trade in Sierra Leonean diamonds and of the arms trade networks supplying the RUF.

In late December 2000, the Panel of Experts on Sierra Leone Diamonds and Arms released a report on its findings.[89] The report described in detail a system of trade, linked to international criminal networks, that has enabled the export of diamonds from Sierra Leone and the export of arms into the country, despite U.N. sanctions prohibiting these activities. The report named many individuals, businesses, and countries that it found to be complicit in these activities, and described how the RUF has exported diamonds from Sierra Leone. It focused much of its attention on the role of Liberia and several other African countries in enabling such prohibited trade but also underlined the failure of European countries to more strongly regulate arms brokering and rough diamond trading. It noted that lax regulation of trade between European countries, including Belgium, Switzerland, and the United Kingdom (UK), enabled the origin of diamonds from conflict regions to be obscured. The report, which also described systemic weaknesses in West Africa aircraft control systems, included many detailed recommendations on how to prevent prohibited arms and diamond trading. It also noted a high degree of overlap between the work and findings of the Sierra Leone and the Angolan sanctions committees and suggested that the United Nations create a permanent sanctions monitoring and compliance body.

---

[89] S/2000/1195. Also see footnote __, above.

Certification System[90]

The Sierra Leone certificate system was not initially accepted by the Sanctions Committee, which met on August 9, 2000 and September 29 to consider in detail the GOSL plan for compliance with U.N. Resolution 1306. On October 6, the Sanctions Committee effectively accepted the GOSL export control plan, and recommended to the Security Council that the diamonds exported under the GOSL regime be exempted from the sanctions imposed by Resolution 1306. On October 12, the GOSL reportedly lifted the ban on legal diamond exports.

The certification system, which the GOSL developed in consultation with experts from the governments of the United States, the United Kingdom, Israeli experts, and with technical assistance from the Diamond High Council, is reported to operate as follows:

- Each parcel is to be accompanied by a certificate of origin (COO) labeled for identification with a randomly chosen serial number printed on tamper-proof watermarked paper.

- Each COO is to be matched by an import confirmation certificate and a numbered label, both originating as perforated sections of the original certificate, and incorporating the same security measures. Under the system, the import confirmation certificate is detached from the COO upon arrival of a diamond parcel at its destination by the importing authorities and is returned to the exporting authority in Sierra Leone. The numbered label is placed as a legal seal upon the parcel being shipped to prevent tampering with the package en route.

- The certificates, printed in the United Kingdom and paid for by the GOSL, will be issued by the Sierra Leonean Gold and Diamond Office. Their cost will be recouped by exporting firms upon issuance through a payment of a certification fee.

- Each parcel is also to be accompanied by a GOSL valuation form, describing the consignment: by weight and value, and describing associated export duties and other shipment identification

---

[90] The proposed certification system is described in *Presentation by the Government of the Republic of Sierra Leone to the United Nations Sanctions Committee*; interview with USAID official, Sept. 22, 2000; Office of Transition Initiatives, USAID, *Status Report and Proposed Actions: Technical Assistance to the Government of Sierra Leone for Development and Implementation of a New Mining and Export Regime for Diamonds*, September 1, 2000.

information. The valuation form information is reflected upon the reverse of each certificate of origin.

- Each parcel is also to be accompanied by a commercial invoice, containing information and in a format that conforms to international trade practice and has been approved by the Bank of Sierra Leone.

The GOSL export regime also incorporates a separate security system that is meant to augment and add redundancy to the COO system. It consists of digital photographs of the contents of each certified export parcel accompanied by a digital record reflecting the same identifying export information found in the COO documents described above. This digital information is transmitted separately to the recipient export authorities upon the dispatch of each parcel shipment. It comprises an independent, parallel means of verification for each shipment.

The COO regime provides for the seizure of parcels that have been found to have been tampered with during shipment, which will be undertaken by a limited number of exporting firms. Each participating firm must be licensed in Sierra Leone and hold membership in diamond exchanges recognized by the World Federation of Diamond Bourses or the International Diamond Manufactures Association, two major industry groups. Each participating firm is also required to train and employ as a valuer at least one Sierra Leonean. Each must also buy and sell its diamonds in accordance with a revised export-banking regime, currently under development by the GOSL, aimed at creating a secure and efficient fiscal and financial system. The envisioned regulatory system will reportedly allow for the purchase of diamonds using hard currency, the import and export of which must be undertaken in transactions channeled through accounts held in banks accredited by the Bank of Sierra Leone. Such accounts will reportedly allow for open market convertibility of hard currency, the establishment of which is aimed at preventing black market currency arbitrage and illegal expatriation of foreign currency and diamond sales profit.

Broad Context of Regulatory Reform

The nascent COO export system, which revises and improves upon an existing GOSL diamond trading and export legal regime, is meant both to comply with the U.N. Resolution 1306 and to prevent the export of conflict diamonds through improved regulation of exports. It is also meant, however, to bring about conformity with international norms of commerce and to foster reform of the domestic trading system in order to create a diamond industry based on open market principles that will better distribute diamond-related wealth.

Ensuring legal consistency with international trade practices is meant to regularize the diamond trade, at least with regard to Sierra Leonean diamonds, by making it subject to integration with a broad body of laws governing trade. Within Sierra Leone, the export system is meant to foster broad-based reform aimed at creating a transparent diamond industry built upon positive financial incentives and inclusive social participation that will, in turn, be amenable to regulation by a combination of laws and self-interest among participants in an open market. These reform efforts – of which free floating currency exchange is an integral part – aim at removing structural incentives, such as government-controlled rates of exchange, that create opportunities for illicit activities (e.g., smuggling, black market currency transactions and expatriation of diamond earnings). Likewise, they are meant to remove artificially high barriers to entry into legitimate trade, such as unreasonably high tariffs and licensing fees, and opportunities for exploiting unequal terms of trade among producers and traders, such as lack of access to credit, price differentials between legal and illicit trading.[91]

Future Prospects

Although action at the international level is being taken to curtail the trade in conflict diamonds, an effective and conclusive solution to the problem is unlikely in the short term. The challenge is two-fold:

- The creation of international consensus, backed up by legislative and administrative action to create a global diamond trade regime.

- The need to strengthen the domestic capacity of Sierra Leone, Angola, and the DRC to effectively regulate their economies.

---

[91] The relation between structural market distortions and disincentives and the rise of illicit and policy strategies are described in detail in "Appendix. Guidelines on the mining and marketing of diamonds in Sierra Leone." *Presentation by the government of the Republic of Sierra Leone to the United Nations sanctions committee*, July 31, 2000. virtually the same document, which proposes a wide range of market-based reform solutions, is available at USAID, *Diamonds and Armed Conflict in Sierra Leone*.

Progress at the international and national levels is on-going, but at the root of the problem is armed conflict, poverty, and varying degrees of institutional incapacity aggravated by corruption. These problems require long-term, comprehensive socio-economic policy solutions, of which conflict diamonds are merely a component. However, the many current policy initiatives described in this report are the basis for incremental progress in dealing with the problem of conflict diamonds.

The Angolan and Sierra Leonean diamond export control systems are administratively and procedurally robust in concept. They are workable policy tools that are presently being implemented, and represent a first concrete step toward reducing conflict diamond exports. They also represent a basic building block in efforts to promote the conformance of the international diamond trade with standard international trade mechanisms and national laws.

The DRC certification or origin process is less robust. It relies upon the IDI deal, which incorporates a range of provisions for ensuring that diamonds exported from the DRC originate in territory controlled by the government. The IDI deal, however, may not endure, according to industry observers, and the DRC government does not appear to have formulated a clear-cut policy on diamond export controls apart from its announcement of the IDI deal. Both the Congolese and the Angolan single buyer policies could also lead to increased smuggling, if sub-market prices are paid to local diamond dealers, as has been reported in Angola, or if the single buyer is not willing or able to purchase the available supply, as has been reported in the Congo.

The international diamond trade has historically been self-regulating and, in many respects, has been exempt from many standard trade laws governing trade in goods. Illicit trading in diamonds exploits these aspects of the legitimate trade, and these factors, along with the fungible nature of diamonds, has made the prosecution of crimes associated with illicit diamond trading difficult to detect and prosecute. The application of standard regulatory norms of trade in diamonds, such as requiring certificates of origin during export, is aimed at removing such exemptions of practice. It is also meant to establish a body of standard export, trade, and tariff documentation, related to diamond transactions, that can be audited and tracked. Such documentation, in turn, may in the future also enable larger bodies of national and international law and legal precedent (including punitive measures, standards of evidence, and other legal tools and remedies) to be applied to illicit diamond trading. In addition, it facilitates the possible future application to diamond trading of trade system incentives, such as the application of tariff reductions under various preferential tariff agreements.

Regardless of the effectiveness of export control policy planning, it remains possible that conflict diamonds might enter into legitimate commerce –albeit likely at reduced prices – through the domestic trading systems of Sierra Leone, the DRC, and Angola. This possibility remains because these countries, which remain in conflict, possess weak regulatory and administrative control over much of their territories and economies.[92] Both the Angolan and Sierra Leonean plans take cognizance of these challenges, and incorporate measures to better regulate their respective domestic industries. The GOSL diamond industry plans also incorporate measures for institutionalizing open market competition and transparency in diamond trading and for broadening of participation in the diamond sector through the increased access to credit and the development of mining and trading cooperatives. These measures, if implemented, are likely to have the greatest impact on reform by legitimizing and spreading the wealth created by diamonds.

Selected Web Sites

Bibliographic Note

Key United Nations documents that relate to conflict diamonds, such as Sanctions Committee reports, are available from the Global Policy Forum site (see below). The Global Policy Forum site also includes reports and proposals produced by non-governmental organizations and the diamond industry.

Amnesty International-USA, *Clean Diamond Act.* [http://www.amnesty usa.org/diamonds/]
Brackenbury, Andrew, "Geopolitics: Gem Warfare", *Geographical*, January 2001. [http://www.geographical.co.uk/geographical/features/jan_2001_diamonds.html].
De Beers, *Conflict Diamonds.* [http://www.debeerscanada.com/conflict/].

---

[92] Anecdotal information suggests that Sierra Leone may currently be facing difficulties implementing the domestic aspects of its new diamond control policy. See Norimitsu Onishi, "Africa Diamond Hub Defies Smuggling Rules," *New York Times*, January 2, 2001, pages A1 and A8.

Department of Minerals and Energy, Mineral Development Branch [South Africa], *Technical Forum on Diamonds*, Kimberly, South Africa, May 11-12, 2000. [http://www.dme.gov.za/minerals/forumvolweb.html].

Diamond High Council (HRD), *Conflict Diamonds: Analyses, Actions, Solutions*. [http://www.conflictdiamonds.com].

Diamond Trade Net, *Conflict Diamond Forum*. [http://www.diamonds.net/conflictdiamonds/]

Fatal Transactions, *International Diamond Campaign* [non-governmental coalition]. [http://www.niza.nl/uk/campaigns/diamonds/index.html].

Global Policy Forum, Issue: Diamonds in Conflict. [http://www.globalpolicy.org/security/issues/diamond/index.htm].

Global Witness, *A Rough Trade: The Role of Companies and Governments in the Angolan Conflict*, December 1998. [http://www.oneworld.org/globalwitness/reports/Angola/cover.htm]

--*Conflict Diamonds: Possibilities for the Identification, Certification and Control of Diamonds*, May 2000. [http://www.oneworld.org/globalwitness/reports/conflict/cover.htm].

Goreux, Louis, "Conflict Diamonds." Africa Region Working Paper Series No. 13. World Bank. February 2001. [http://www.worldbank.org/afr/wps/wp13.htm].

Guerrera, Francesco, Mark Huband, Andrew Parker, and Sathnam Sangera, "Fatal Transactions: An Investigation into the Illicit Diamond Trade," *Financial Times*, July 12, 2000. [http://www.ft.com/diamonds/].

Harden, Blaine, "African's Diamond Wars," *New York Times*, undated. [http://www.nytimes.com/library/world/Africa/040600africa-diamonds.html].

Partnership Africa Canada, *The Heart of the Matter: Sierra Leone, Diamonds and Human Security*, January 2000. [http://www.partnershipafricacanada.org/englich/esierra.html].

Physicians for Human Rights, *Sierra Leone*. [http://www.phrusa.org/campaigns/sierra_leone/index.html].

Terpstra, Arjan and Gerbina van den Hurk. EU Control of Diamond EU Control of Diamond Imports from African Countries in Conflict. Report of the European Union Expert Meeting. Fatal Transactions. September 25, 2001. [http://www.niza.nl/uk/campaigns/diamonds/index.html].

United Nations, Department of Public Information with the Sanctions Branch, Security Council Affairs Division, Department of Political Affairs, *Conflict Diamonds: Sanctions and War*. [http://www.un.org/peach/Africa/Diamond.html].

U.S. Agency for International Development, Bureau of Humanitarian Response, Office of Transition Initiatives, *Diamonds and Armed Conflict in Sierra Leone: Proposal for Implementation of a New Diamond Policy and Operations*, May 8, 2000. [http://www.usaid.gov/hum_response/oti/country/sleone/diamonds.html].

U.S. Department of State, Mission to Botswana, *Links to Information on "Conflict Diamonds."* [http://www.usembassy.state.gov/posts/bc1/wwwhdiam.html].

U.S. Department of State, Office of the Spokesman, *Fact Sheet: U.S. Initiatives on "Conflict Diamonds,"* May 23, 2000.

U.S. Geological Survey, *Africa and the Middle East: Angola.* [http://minerals.usgs.gov/minerals/pubs/country/Africa.html#ao].

U.S. Geological Survey, *African and the Middle East: Congo Kinshasa.* [http://minerals.usgs.gov/minerals/pubs/country/Africa.html#cg.]

U.S. Geological Survey, *Africa and the Middle East: Sierra Leone.* [http://minerals.usgs.gov/minerals/pubs/country/Africa.html#sl].

U.S. Geological Survey, *Minerals Information.* [http://minerals.usgs.gov/minerals/]

U.S. House of Representatives, Committee on Ways and Means, Subcommittee on Trade, *Hearing on Trade in African Diamonds* [Witness List and Testimony], 106th Congress, September 13, 2000. [http://waysandmeans.house.gov/trade/106cong/tr-23wit.htm].

Vander Stychele, Myriam. Conflict Diamonds: Crossing European Borders? A Case Study of Belgium, the United Kingdom and the Netherlands. Fatal Transactions. August 2001. [http://www.niza.nl/uk/campaigns/diamonds/index.html].

World Diamond Council. [http://www.worlddiamondcouncil.com/].

Chapter 2

DIAMOND-RELATED AFRICAN CONFLICTS: A FACT SHEET

Nicholas Cook and Jessica Merrow

INTRODUCTION

This chapter summarizes major demographic and spending trends that characterize the on-going conflicts in the Democratic Republic of the Congo (DRC), Sierra Leone, and Angola.[1] In each of these conflicts, contention over the control of mineral wealth, particularly diamonds – and other natural resources is regarded as a significant factor fueling hostilities. This fact sheet presents data on refugees, internally displaced persons (IDPs), deaths, child soldiers approximate percentage of territory held by anti-government rebels, military spending and international humanitarian spending.

GENERAL CAVEAT ON CONFLICT-RELATED STATISTICS IN AFRICA

In general, conflict-related statistics – including those contained in this fact sheet – should be treated as rough estimates. This is particularly true for

---

[1] For background on the relation between diamonds and conflict in Africa, see CRS Report RL30751, *Diamonds and Conflict: Policy Proposals and Background.*

conflicts in Africa, where many governments lack adequate resources to undertake statistical profiling, even in peaceful settings. Many conflict-related figures are derived from statistical projections based on aging data sets (past censuses, etc.), small rapid assessment survey samples, and variable assumptions about demographic profiles resulting from "normal" levels of poverty, disease, access to health care, and the like. In many cases, it is nearly impossible to definitively differentiate between the effects of conflict, natural disaster, disease, and general poverty. The figures below are derived from multiple information sources and from studies employing a wide variety of research methodologies. With the above qualifications, the following facts, figures and data sources relating to the three current diamond-related African conflicts are given:

SIERRA LEONE

Refugees

Approximately 490,000 as of September 2000 (mostly in Guinea, Liberia and Ivory Coast).[2]

Internally Displace Persons (IDPs)

500,000 to over one million. 300,000 additional displacements due to May 2000 upsurge in fighting.[3]

Deaths

Estimates for total deaths related to conflict since 1991 range between 20,000 and 50,000.[4]

---

[2] U.S. Agency For International Development (USAID), *Sierra Leone – Complex Emergency Fact Sheet #2*, September 13, 2000.
[3] *Ibid.*
[4] USAID, *Sierra Leone – Fact Sheet #2* and Project Ploughshare, *Armed Conflicts Report 2000* [http://www.ploughshares.ca/content/ACR/ACR00/ACR00.html]. Systematic and repeated incidents involving the mutilation, abduction, rape, and other human rights abuses of civilians have been documented by international human rights groups and United Nations (U.N.) agencies. Aggregate totals for such atrocities, however, are not available.

Child Soldiers

5,000 to 5,400, in direct combat roles. 5,000 or more used in combat support roles.[5]

Approximate Rebel-held Territory

One-third to one-half of the national territory (multiple press accounts).

Military Spending

$11 million (1999); $46 million (FY1996/1997);[6] $476.7 million appropriated by U.N. General Assembly for the United Nations Mission in Sierra Leone (UNAMSIL), June 30, 2001 through July 1, 2001.[7]

Humanitarian Spending

$79 million estimated need for 2001.[8]

ANGOLA

Refugees

Over 340,000 in neighboring countries (November 2000 estimate).[9]

---

[5] Amnesty International, *Sierra Leone: Childhood – A Casualty of Conflict*, August 31, 2000 [http://www.amnesty.ca/library/afr5106900.htm]; and World Bank Administered Multi-Donor Trust Fund for the Sierra Leone Disarmament, Demobilization and Reintegration Program *Progress Report #3*, June 30, 2000, also available on the Internet and the World Bank's web site [http://www.worldbank.org/afr/afth2/crrp/crrp_mdtfreport3.html].

[6] International Institute for Strategic Studies, *The Military Balance 2000-2001*; Central Intelligence Agency, *The World Factbook 2000* [http://www.odci.gov/cia/publications/factbook].

[7] *Eighth report of the Secretary-General on the United Nations Million in Sierra Leone* [S/2000/1199], December 15, 2000, page 9.

[8] UN Office for the Coordination of Humanitarian Affairs (OCHA), *UN Consolidated Inter-Agency Appeal for Sierra Leone 2001*, November 14, 2000. In 2000, a Consolidated Inter-Agency Appeal for $71 million yielded a 65.4% donor response.

[9] Norwegian Refugee Council Global IDP Database *Angola: Profile Summary*.

Internally Displaced Persons (IDPs)

Approximately 1 million newly displaced. 2000: Approximately 338,000 newly displaced; 2.7 million total displaced as of late July 2000.[10]

Deaths

Minimum of 650,000 conflict-related deaths, 1974 through 1999.[11]

Child Soldiers

5,000-7,000 estimated as of 1997; figure could be substantially higher (various sources).

Approximate Rebel-held Territory

Varies. UNITA guerrilla tactics make much of the country insecure; UNITA has been active in most parts of Angola during the past year, and is believed to effectively control a large area adjacent to the Zambian border.

Military Spending

$1.005 billion in 1999, estimated; $1.2 billion (FY1997/1998).[12]

Humanitarian Spending

$202 million estimated need for 2001.[13]

---

[10] OCHA, *UN Consolidated Inter-Agency Appeal for Angola 2001*, November 9, 2000 and *Report of the Secretary-General on the United Nations Office in Angola* (S/2000/977), October 10, 2000.
[11] Heidelberg Institute of International Conflict Research, *Database KOSIMO 1945-1999*; Energy Information Administration, U.S. Department of Energy, *Country Analysis Briefs, Angola*.
[12] International Institute for Strategic Studies, *The Military Balance 2000-2001*; Central Intelligence Agency, *The World Factbook 2000* [http://www.odci.gov/cia/publications/factbook].
[13] OCHA, *UN Consolidated Inter-Agency Appeal for Angola 2001*, November 9, 2000. In 2000, a Consolidated Inter-Agency Appeal for $260.6 million yielded a 52.5% donor response.

DEMOCRATIC REPUBLIC OF THE CONGO (DRC)

Refugees

162,000 in Uganda, Tanzania and Zambia (January 2001); estimated 310,000 in these and other neighboring states (November 2000); the DRC hosts 335,800 refugees from neighboring countries (December 2000).[14]

Internally Displaced Persons (IDPs)

1.8 million (November 2000; see footnote

Deaths

In eastern DRC: 200,000 due directly to acts of violence; a total of 1.7 million direct and indirect war-related deaths were estimated. This estimate covers August 1998 through May 2000.[15]

Child Soldiers

Concise figures not available; child recruitment is frequently reported as being common and widespread; 10,000 child soldiers were estimated to have fought on behalf of the Alliance of Democratic Forces for Liberation in 1997.[16]

Approximate Rebel-held Territory

50 to 60% (multiple press accounts).

---

[14] U.N. High Commissioner for Refugees (UNHCR), *Democratic Republic of the Congo in Short*, [2001 Global Appeal], December 2000; UNHCR, *DR Congo: The Impact of Refugees*, January 2001.

[15] International Rescue Committee, Mortality Study, Eastern Democratic Republic of Congo, May 2000, [http.www.theirc.org/mortality.cfm].

[16] International Coalition to Stop the Use of Child Soldiers, *The Use of Children as Soldiers in Africa: A Country Analysis of Child Recruitment and Participation in Armed Conflict*, August 14, 2000.

Military Spending

$400 million (1999, estimated); $250 million (FY1997);[17] Commitment authority of U.N. Organizational Mission in the Democratic Republic of the Congo (MONUC); $141.3 million gross July 2000 through June 2001.[18]

Humanitarian Spending

$139.4 million estimated need for 2001.[19]

---

[17] International Institute for Strategic Studies, *The Military Balance 2000-2001*; Central Intelligence Agency, *The World Factbook 2000* [http://www.odci.gov/cia/publications/factbook].

[18] U.N., Democratic Republic of the Congo – MONUC Facts and Figures; available online at [http://www.un.org/Depts/dpko/monuc/monucF.htm].

[19] OCHA, *U.N. Consolidated Inter-Agency Appeal for Democratic Republic of the Congo 2001*, November 8, 2000. In 2000, a Consolidated Inter-Agency Appeal yielded $77 million in donor assistance.

Chapter 3

DIAMONDS OF DEATH

Ken Silverstein[*]

For much of last year, the diamond industry was rocked by a wave of bad publicity concerning "conflict diamonds"-smuggled gems whose sale allows African governments and rebel groups to finance their wars. In addition to several United Nations reports and hearings in the US Congress on conflict diamonds, the media devoted considerable space to the topic. Particularly damaging was a PrimeTime Live segment that included dramatic footage from Sierra Leone, where rebels known as the Revolutionary United Front-best known for cutting off the limbs of civilians who oppose it-fund themselves primarily through diamond sales. Meanwhile, groups such as Global Witness, World Vision, Physicians for Human Rights and Amnesty International threatened to launch a consumer boycott until the industry changed its buying practices so as to insure that conflict diamonds are eliminated from international markets.

Fearful that diamonds might become to this decade what fur was to the last, industry leaders in the United States (such as Lazare Kaplan International) and abroad (especially De Beers of South Africa) vowed to take action. To address the issue of conflict diamonds-which account for 4 percent of the world's $7-billion-a-year trade according to industry and at least 15 percent according to human rights groups-the companies formed the

---

[*] *Ken Silverstein is an investigative reporter based in Washington, DC. Research support was provided by the Investigative Fund of the Nation Institute.*

World Diamond Council last July. The WDC pledged to support the campaign for reform, including efforts to halt imports of conflict diamonds into the critical US market, where about two-thirds of diamonds are sold.

Instead, diamond companies and trade groups have launched a lavishly funded lobbying and public relations campaign aimed at burnishing the industry's reputation while protecting its profits. The centerpiece of the industry effort is a bill called the Conflict Diamonds Act of 2001, which was written by one of the Beltway's most well-connected lobby shops. Representative Tony Hall, an Ohio Democrat who has spoken at rallies against conflict diamonds at jewelry stores in New York City and Chevy Chase, Maryland, calls the act "mushy, fishy and full of loopholes." "The industry is completely untrustworthy," says Hall, who along with Republican Representative Frank Wolf and Democratic Representative Cynthia McKinney has introduced an opposing bill to crack down on conflict diamonds. "The fact that they've hired so many of the top lobbyists in town shows that they don't want any serious legislation to get passed."

During the cold war, dictators and guerrillas in Africa could turn to one of the superpowers for financial support. Now rebel groups need to raise their own money to buy weapons and pay for their wars. In Africa the source of that money is invariably diamonds, which are small, easy to smuggle and hugely lucrative. The three primary sources of conflict diamonds are Angola, Sierra Leone and the Democratic Republic of Congo (formerly Zaire). In the first, Jonas Savimbi's UNITA rebels, supported until the early 1990s by cash and matériel from the CIA, have raised almost $4 billion from diamond sales during the past decade. In Sierra Leone, the Revolutionary United Front has smuggled out at least $630 million in diamonds through Liberia, in exchange for support and weaponry from Charles Taylor, the corrupt warlord who heads that nation. To pay for its war against rebel groups and their foreign allies, the government of the Democratic Republic of Congo formed a diamond joint venture with the armed forces in neighboring Zimbabwe. The combined death toll from fighting in the countries named above is more than 2 million. About a quarter of the casualties are in Angola, where four-fifths of the country's population lives in poverty, average life expectancy is 42 and more than 2 million people are internal refugees.

Countries that launder diamonds smuggled out of war zones also profit from the conflict diamond trade. The big players here, in addition to Liberia, are the Ivory Coast, Guinea, Gambia, Togo, Burkina Faso, Zimbabwe and Ukraine (the last of which is accused by the UN of bartering arms for diamonds from UNITA). Profiting, too, are diamond finishing and polishing centers, notably Belgium and Israel. "Why should Antwerp and Tel Aviv be built on the limbs and backs of Freetown, Luanda and Kinshasa?" Representative McKinney asks.

Of course, the other big beneficiaries of the conflict diamond trade are the corporations that control the industry. De Beers, which buys about two-thirds of the world's diamonds, reported an 83 percent increase in profits for 1999. Though it claims to no longer buy conflict diamonds, just a few years ago De Beers publicly acknowledged-indeed, boasted-that it was purchasing most of UNITA's output in order to prevent a glut on the world market that would drive prices down.

Lazare Kaplan International, America's largest diamond cutter, is headed by Maurice Tempelsman, whose initial foray into diamonds came in Zaire, where he was a close friend of the tyrant Mobutu Sese Seko. In the 1960s Tempelsman hired as his business agent the CIA station chief in Kinshasa, Larry Devlin, who helped put Mobutu in power and afterward served as his personal adviser. Tempelsman is best known as Jackie Onassis's longtime companion, but he's also a prominent player in Washington. He's doled out about $500,000 to the Democratic Party in the past decade, and during the Clinton years he was twice invited to state dinners and was one of a dozen executives who went on the President's 1998 trip to South Africa.

Lazare Kaplan has a long-term agreement to buy $100 million worth of gems per year from ALROSA, the Russian diamond monopoly that is rife with corruption and a fierce opponent of efforts to restrict the trade in conflict diamonds. In the mid-1990s the US Export-Import Bank issued a $62 million loan guarantee to underwrite ALROSA's acquisition of mining equipment from Caterpillar. In March, the Ex-Im confirmed that it has approved a new, $200 million loan guarantee that will benefit Lazare Kaplan's diamond-cutting factories in Russia. In January Lazare Kaplan announced that last year's second-quarter sales of polished diamonds increased by 108 percent.

Retailers such as Cartier and Tiffany are also keenly following the political maneuvering on conflict diamonds. Mustered in the Jewelers of America, they form a potent lobbying force, especially with House Republicans.

When the conflict diamond issue first hit the public's radar screen, the industry argued that there wasn't much it could do because it's almost impossible to determine where a diamond was originally mined. Hence, said the gem companies, any effort to eliminate conflict diamonds would wreck the entire trade, thereby devastating legitimate producers like South Africa,

Namibia and Botswana, not to mention thousands of small businesses in the United States and Europe.

Last May, when bad press was starting to take a political toll, a South African initiative brought together governments, industry and critics. What emerged from that meeting, held in the South African city of Kimberley, was a call for a broad system of reform called Rough Controls. Under Rough Controls, diamond-mining countries are to ship stones in tamper-proof containers that will include country-of-origin documentation. All importing countries, including cutting and polishing centers-major ones include Belgium, Israel, Russia and India-will reject stones that can't be certified as coming from countries that have Rough Controls. When the World Diamond Council was founded two months after the Kimberley meeting, it embraced Rough Controls and said that anyone found dealing in conflict diamonds would be permanently banned from the industry.

Buoyed by these shows of good faith, human rights activists and members of Congress sought to put words into action. After months of negotiations with industry representatives, a deal was struck that all the players, including the Clinton Administration, agreed to. Last fall, Hall and his allies attached a rider to an appropriations bill for the Commerce, Justice and State departments that would have prohibited the importation of diamonds from conflict-diamond countries. In the Senate, Judd Gregg, a conservative Republican from New Hampshire-who has compared buying conflict diamonds to purchasing goods made by Nazi Germany-threatened to cut off funding to US Customs if the rider didn't go through with the bill.

Then on December 8, when it appeared that there were no serious obstacles to passage, the World Diamond Council invited activists to a breakfast meeting in Washington and announced that it was withdrawing support for Hall's bill. According to the WDC's executive director, Matthew Runci, the law was badly drafted (this was a surprise, since it was put together by the legislative counsel of the House and was based on a memo from the Jewelers of America) and that industry had therefore decided to write its own legislation. Soon afterward, Hall's rider was stripped from the appropriations bill, which then sailed through Congress. No one knows for sure exactly who did the dirty work, but the move clearly had the blessing of the House Republican leadership.

The WDC recently unveiled its bill-it will likely have been introduced by the time this story goes to press-which was drafted by two lawyers at Akin, Gump, Strauss, Hauer & Feld. Akin, Gump's most notable door openers during the Clinton years were Vernon Jordan and Robert Strauss, but the firm is equally well connected to the GOP. Nine officials from the lobby shop served as members of the Bush Administration's Transition Advisory Teams, including Bill Paxon, the former Congressman from New York, who remains close to the House Republican leadership.

The WDC's bill, the Conflict Diamonds Act of 2001, states that no diamonds are to be imported into the United States from countries that are not on an approved list that the Treasury Department will issue later this year. The problem is that the standards for making the list are incredibly weak. A "cooperating country" is defined by the bill as one that is "negotiating in good faith to develop an acceptable international agreement" or "acting in good faith" to develop a unilateral certification system. "With these flimsy guidelines, virtually any nation can make the [Treasury Department's] good-guy list," says Holly Burkhalter of Physicians for Human Rights, coordinator of the eighty-member coalition working to support the reform movement in Congress.

The WDC's draft bill exempted diamond jewelry from the import ban. Hence, a $20 setting could turn an illegal diamond into a perfectly allowable necklace. In response to criticism from NGOs, Akin, Gump inserted a clause that allows the President to ban jewelry imports if he so chooses. Another section of the WDC bill permits violators of the law to escape prosecution if they import contraband "through inadvertence, or by reason of clerical error or other mistake of fact." "It's a trade lawyer's dream," Deborah DeYoung, an aide to Hall, says scornfully of the WDC's legislation. "It won't cut the flow of conflict diamonds, and there's no incentive for countries to take serious action."

Eli Izhakoff, chairman of the WDC, says that jewelry was exempted because the problem lies with rough stones and that the Akin, Gump bill will insure that only clean diamonds find their way to polishing and finishing centers. "We're looking to make the law efficient and practical," he said by telephone from New York. "If you start making it too costly, it's not going to happen." And why, I asked, did the council draw up its own bill instead of asking for changes to legislation introduced last year that it had pledged to support? "I would rather not have [gone to Akin, Gump]," he says. "It cost a lot of money, and I'm very cheap that way. But this will affect the worldwide industry, and it had to be prepared properly."

Over at Akin, Gump, Warren Connelly, head of the firm's international trade practice group and co-author of the WDC's legislation, says industry is "amenable to discussing" a flat ban on jewelry, but that inserting such a provision is complicated. "We have a concern about how you trace a diamond into a piece of jewelry and putting too heavy a burden on jewelers in enforcing that," he says.

Neither Connelly nor Izhakoff will yet say who the WDC is working with on Capitol Hill, but to make sure that its legislation gets through Congress-and to insure that a new bill introduced by Hall and his allies doesn't-the industry has hired a number of prominent Beltway influence-peddlers. In addition to Akin, Gump, the World Diamond Council is using two big PR firms, Powell Tate-headed by Jody Powell, the former spokesman for President Jimmy Carter, and Sheila Tate, who performed the same job for Nancy Reagan-and Shandwick Associates, a firm that specializes in corporate "grassroots" campaigns. The latter has represented US clients like Monsanto, Ciba-Geigy, Procter & Gamble and Georgia-Pacific. Its clients abroad, according to the newsletter PR Watch, include Timberlands, a New Zealand government-owned logging firm that required help in spinning its pillage of rainforest lands; and Royal Dutch/Shell, which hired Shandwick to counter protests against its operations in Nigeria, where it has closely collaborated with the military.

And that's just the start of the industry's hiring spree. On the payroll of Lazare Kaplan is Ted Sorensen, a former top adviser to President John F. Kennedy, and Kate McAuliffe, a former aide to House minority leader Richard Gephardt, both from the New York law firm of Paul, Weiss, Rifkind, Wharton & Garrison. ("As an active figure in the Democratic Party, he has participated in nine of the last 11 Democratic Party National Conventions and is experienced in the ways of Washington," says the bio of Sorensen on his firm's website.) Tiffany has signed up the blue-chip firm of Cassidy & Associates, which has deployed Christy Evans, previously with the House Republican Conference, and Dan Tate Jr., a former lobbyist for the Clinton White House. The Jewelers of America has turned for help to Haake and Associates, where another revolving-door alumnus, Timothy Haake, is handling the account.

Foreign nations that would be hurt by efforts to ban conflict diamond imports have also shelled out big bucks to fight off tough legislation. Herman Cohen, a Republican foreign policy guru who previously served as Assistant Secretary of State for African Affairs in the first Bush Administration, is lobbying on behalf of the Democratic Republic of Congo and Burkina Faso. Botswana has a good record on conflict diamonds but is nervous about Hall's approach. It has hired Hill and Knowlton to work the Hill, especially the thirty-seven-member Congressional Black Caucus. Even Liberia has bought its own hired gun in the form of Ken Yates of Jefferson Waterman International, a firm whose past clients include Burma's military rulers. (In a bold display of principle, Jefferson Waterman quit when the Burmese fell behind on their payments.)

The industry's phalanx of lobbyists have slowly fanned out across Washington. Before George W. Bush's inauguration, they met with officials at the White House, the State Department and the Treasury Department. To garner further support for the WDC bill, jewelers and jewelry-store owners visited dozens of Congressional offices on February 27, which industry set as the D-day for its lobbying drive.

On the other side, Hall has just reintroduced a version of his bill, which has 107 co-sponsors, including all but seven members of the Black Caucus. It's similar to last year's measure but includes a number of new clauses, including one that bars the Ex-Im from making loans to countries that are not seeking to end the trade in conflict diamonds-what one Congressional staffer dubs the Maurice Tempelsman Memorial Provision. Given the firepower deployed by industry, however, it's not at all clear that the bill can make it through Congress-which from the industry's point of view is almost as good as winning passage of its own bill. "We're going to face a huge uphill battle," Burkhalter concedes. "There's no way to pass tough legislation without industry support, and you don't get that support by holding hands. We need grassroots action."

And as Washington debates the matter and the industry seeks half-measures, warlords and guerrillas ring up millions of dollars a day in the sale of conflict diamonds.

Chapter 4

CONFLICT DIAMONDS: SANCTIONS AND WAR[*]

Anna Frangipani Campino

GENERAL ASSEMBLY ADOPTS RESOLUTION ON "CONFLICT DIAMONDS"

Crucial Issue in Fuelling Wars

On 1 December 2000, the United Nations General Assembly adopted, unanimously, a resolution on the role of diamonds in fuelling conflict, breaking the link between the illicit transaction of rough diamonds and armed conflict, as a contribution to prevention and settlement of conflicts (A/RES/55/56). In taking up this agenda item, the General Assembly recognized that conflict diamonds are a crucial factor in prolonging brutal wars in parts of Africa, and underscored that legitimate diamonds contribute to prosperity and development elsewhere on the continent. In Angola and Sierra Leone, conflict diamonds continue to fund the rebel groups, the National Union for the Total Independence of Angola (UNITA) and the Revolutionary United Front (RUF), both of which are acting in

---

[*] Excerpted from the United Nations Department of Public Information website: http://www.un.org/peace/africa.

contravention of the international community's objectives of restoring peace in the two countries.

What is a Conflict Diamond?

Conflict diamonds are diamonds that originate from areas controlled by forces or factions opposed to legitimate and internationally recognized governments, and are used to fund military action in opposition to those governments, or in contravention of the decisions of the Security Council.

> "It has been said that war is the price of peace... Angola and Sierra Leone have already paid too much. Let them live a better life."
> ***Ambassador Juan Larrain, Chairman of the Monitoring Mechanism on sanctions against UNITA.***

How can a Conflict Diamond be Distinguished from a Legitimate Diamond?

A well-structured 'Certificate of Origin' regime can be an effective way of ensuring that only legitimate diamonds -- that is, those from government-controlled areas -- reach market. Additional controls by Member States and the diamond industry are needed to ensure that such a regime is effective. These measures might include the standardization of the certificate among diamond exporting countries, transparency, auditing and monitoring of the regime and new legislation against those who fail to comply.

> "Diamonds are forever" it is often said. But lives are not. We must spare people the ordeal of war, mutilations and death for the sake of conflict diamonds."
> ***Martin Chungong Ayafor, Chairman of the Sierra Leone Panel of Experts***

FUELLING WARS

Rough diamond caches have often been used by rebel forces to finance arms purchases and other illegal activities. Neighbouring and other countries can be used as trading and transit grounds for illicit diamonds. Once diamonds are brought to market, their origin is difficult to trace and once polished, they can no longer be identified.

Who Needs to Take Action?

Governments, inter-governmental and non-governmental organizations, diamond traders, financial institutions, arms manufacturers, social and educational institutions and other civil society players need to combine their efforts, demand the strict enforcement of sanctions and encourage real peace. The horrific atrocities in Sierra Leone and the long suffering of the people of Angola have heightened the international community's awareness of the need to cut off sources of funding for the rebels in order to promote lasting peace in those countries; such an opportunity cannot be wasted.

Legitimate Diamonds → Peace → Development

- Controls on conflict diamonds cut off sources of funding for rebels, help shorten wars and prevent their recurrence.

- Peace in diamond producing regions will bring about the potential for economic development and tax revenue for building infrastructure as legitimate mining ventures increase.

The international diamond industry is already taking steps to respond, such as the adoption by the World Diamond Congress, Antwerp, 19 July 2000, of a resolution which, if fully implemented, stands to increase the diamond industry's ability to block conflict diamonds from reaching market. Other efforts include the launching, at the initiative of African diamond-producing countries, of an inclusive, worldwide consultation process of Governments, industry and civil society, referred to as the Kimberly Process, to devise an effective response to the problem of conflict diamonds.

What is the United Nations Doing?

The tragic conflicts in Angola and Sierra Leone, fuelled by illicit diamond smuggling, have already led to action by the Security Council. Under Chapter VII of the United Nations Charter, targeted sanctions have been applied against UNITA in Angola and the Sierra Leone rebels, including a ban on their main source of funding -- illicit diamonds. Diamond sanctions have also been applied against Liberia but are not yet in effect.

Angola

Following UNITA's rejection of the results of the United Nations monitored election in 1992, the Security Council, acting under Chapter VII of the United Nations Charter, adopted resolution 864 of 15 September 1993, imposing an arms embargo along with petroleum sanctions against UNITA and establishing a Sanctions Committee consisting of all the members of the Council to monitor and report on the implementation of the mandatory measures.

This UNITA soldier, here awaiting demobilization, was recruited at age 11. Vila Nova, Angola, 1998. UNICEF / HQ96-008 / Giacomo Pirozzi

Following the signing of the 1994 Lusaka Protocol, UNITA refused to comply with its terms. In response to UNITA's refusal to disarm and implement the Lusaka Protocol, the Security Council adopted resolution 1127 of 28 August 1997, which imposed mandatory travel sanctions on senior UNITA officials and their immediate family members. A year later, the Security Council adopted resolution 1173 of 12 June 1998 and resolution 1176 of 24 June 1998, prohibiting the direct or indirect import from Angola to their territory of all diamonds not controlled through the Certificate of Origin issued by the Government of Angola, as well as imposing financial sanctions on UNITA.

By resolution 1237 of 7 May 1999, the Security Council established an independent Panel of Experts to investigate violations of Security Council sanctions against UNITA. Following the publication of the Panel's report (document S/2000/203), the Security Council adopted resolution 1295 of 18 April 2000, by which the Panel's recommendations were taken up and a "Monitoring Mechanism" was established to collect additional information and investigate any relevant leads regarding sanctions violations, with a view to enhancing the implementation of the measures imposed on UNITA. The five expert members were its Chairman, Ambassador Juan Larrain (Chile), Christine Gordon (United Kingdom), James Manzou (Zimbabwe), Ismaila Seck (Senegal) and Ambassador Lena Sundh (Sweden). The Mechanism submitted its report to the Committee on 20 December 2000 (S/2000/1225). By resolution 1336 (2001), the Security Council extended the mandate of the Monitoring Mechanism for a period of three months. On 20 February 2001, the Security Council held an open meeting to discuss the report of the Monitoring Mechanism.

SIERRA LEONE

In July 1999, following over eight years of civil conflict, negotiations between the Government of Sierra Leone and the Revolutionary United Front led to the signing of the Lome Peace Agreement under which the parties agreed to the cessation of hostilities, disarmament of all combatants and the formation of a government of national unity. The United Nations and the Economic Community of West African States (ECOWAS) helped facilitate the negotiations. In resolution 1270 of 22 October 1999, the Security Council established the United Nations Mission in Sierra Leone (UNAMSIL) to help create the conditions in which the parties could implement the Agreement. Subsequently, the number of personnel were increased and tasks to be carried out by UNAMSIL adjusted by the Council in resolutions 1289 of 7 February 2000 and 1299 of 19 May 2000, making UNAMSIL the largest peacekeeping force currently deployed by the United Nations.

Following international concern at the role played by the illicit diamond trade in fuelling conflict in Sierra Leone, the Security Council adopted resolution 1306 on 5 July 2000 imposing a ban on the direct or indirect import of rough diamonds from Sierra Leone not controlled by the Government of Sierra Leone through a Certificate of Origin regime. An arms embargo and selective travel ban on non-governmental forces were already in effect under resolution 1171 of 5 June 1998.

On 31 July and 1 August 2000, Ambassador Anwarul Karim Chowdhury, Chairman of the Security Council Committee established pursuant to resolution 1132 (1997) concerning Sierra Leone, presided over the first ever exploratory public hearing by the Security Council in New York. The hearing was attended by representatives of interested Member States, regional organizations, non-governmental organizations, the diamond industry and other relevant experts. The hearing exposed the link between the trade in illicit Sierra Leone diamonds and trade in arms and related materiel. The ways and means for developing a sustainable and well-regulated diamond industry in Sierra Leone were also discussed.

As called for by resolution 1306 of 5 July 2000, the Secretary-General, on 2 August 2000, established a Panel of Experts, to collect information on possible violations of the arms embargo and the link between trade in diamonds and trade in arms and related materiel, consider the adequacy of air traffic control systems in the West African region for the purpose of detecting flights suspected of contravening the arms embargo, and report to the Council with observations and recommendations on ways of strengthening the arms and diamond embargoes no later than 31 October 2000. The Chairman of the Panel was Martin Chungong Ayafor (Cameroon). The other members were Atabou Bodian (Senegal), Johan Peleman (Belgium), Harjit Singh Sandhu (India) and Ian Smillie (Canada). The Panel submitted its report to the Security Council on 19 December 2000 (S/2000/1195). On 25 January 2001 the Security Council, at its 4264th meeting, considered the report of the panel of experts.

LIBERIA

Following the findings presented in the Sierra Leone Panel of Experts' report that the illicit trade in diamonds from Sierra Leone could not be conducted without the permission and involvement of the Liberian government officials, and that the Government of Liberia was actively supporting the RUF at the highest levels, the Security Council adopted resolution 1343 of 7 March 2001. By this resolution, a new Sanctions Committee of the Security Council was established, an arms embargo was re-applied and a Panel of Experts was mandated for a period of six months. In addition, the resolution indicated that if the Government of Liberia does not meet the demands specified by the Security Council within two months, all States would be mandated to take the necessary measures to prevent the direct or indirect import of all rough diamonds from Liberia, whether or not such diamonds originated in Liberia, and a selective travel ban would be imposed.

Chapter 5

INTERNATIONAL TRADE: SIGNIFICANT CHALLENGES REMAIN IN DETERRING TRADE IN CONFLICT DIAMONDS

Loren Yager

WHY THIS STUDY WAS DONE

Conflict diamonds are used by rebel movements to finance their military activities, including attempts to undermine or overthrow legitimate governments. These conflicts have created severe humanitarian crises in countries such as Sierra Leone, Angola, and the Democratic Republic of the Congo. An international effort called the Kimberley Process aims to develop a diamond certification scheme to prevent the flow of conflict diamonds. Legislation is also being developed to address U.S. consistency with the Kimberley Process. GAO was asked to assess the challenges associated with deterring trade in conflict diamonds.

WHAT WAS FOUND

The nature of diamonds and the international diamond industry's operations create opportunities for illicit trade, including trade in conflict diamonds. Diamonds are a high-value commodity easily concealed and transported, are mined in remote areas worldwide, and are virtually untraceable to their original source. These factors allow diamonds to be used in lieu of currency in arms deals, money laundering, and other crime. Further, the diamond industry lacks transparency, which limits information about diamond transactions.

U.S. controls over diamond imports generally do not require certification from the country of extraction—just from the country of last import—and thus are not very effective in identifying diamonds from conflict sources. While the United States bans diamonds documented as coming from the National Union for the Total Independence of Angola, the Revolutionary United Front in Sierra Leone, and Liberia—all of which are subject to U.N. sanctions—this does not prevent conflict diamonds shipped to a second country from being mixed into U.S.-destined parcels.

GAO's assessment of the Kimberley Process's proposal for an international diamond certification scheme found it incorporated some elements of accountability. However, the scheme is not based on a risk assessment, and some activities experts deem high risk are subject only to "recommended" controls. Also, the period after rough diamonds enter the first foreign port until the final point of sale is covered by a system of voluntary industry participation and self-regulated monitoring and enforcement. These and other shortcomings provide significant challenges in creating an effective scheme to deter trade in conflict diamonds.

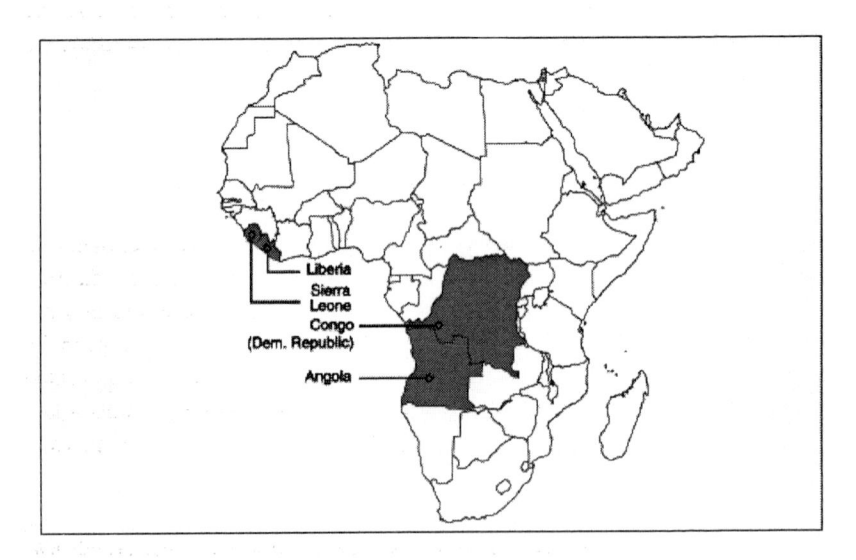

Main Countries Associated With Conflict Diamonds

International Trade: Significant Challenges Remain in Deterring ...

The United Nations General Assembly defines conflict diamonds as rough diamonds used by rebel movements to finance their military activities, including attempts to undermine or overthrow legitimate governments. These conflicts have created severe humanitarian crises in countries such as Sierra Leone, Angola, and the Democratic Republic of the Congo. The United States and much of the international community are trying to sever the link between conflict and diamonds while ensuring that no harm is done to the legitimate diamond industry, which is economically important in many countries. The principal international effort to address these objectives, known as the Kimberley Process, aims to develop and implement an international diamond certification scheme that will deter conflict diamonds from entering the legitimate market. The Kimberley participants, including government, diamond industry, and nongovernmental organization officials, have reported back to the United Nations General Assembly with a proposal they believe provides a good basis for the envisaged scheme.[1] Consistent with the Kimberley Process, the U.S. Congress has legislation pending that would require countries exporting diamonds to the United States to have a system of controls to keep conflict diamonds from entering their stream of commerce.

This chapter discusses (1) how the nature of diamonds and industry operations are conducive to illicit trade; (2) U.S. government controls over diamond imports; and (3) the extent to which the Kimberley Process international diamond certification scheme, in its current form, has the necessary elements to deter trade in conflict diamonds. The observations are based on ongoing work on conflict diamonds.

SUMMARY

The nature of diamonds and the operations of the international diamond industry create opportunities for illicit trade, including trade in conflict diamonds. Diamonds are mined in remote areas around the world and are virtually untraceable back to their original source—two factors that make monitoring diamond flows difficult. Diamonds are also a high-value commodity that is easily concealed and transported. These conditions allow diamonds to be used in lieu of currency in arms deals, money laundering,

---

[1] The proposal was presented in the form of a Kimberley Process Working Document titled *Essential Elements of an International Scheme of Certification for Rough Diamonds, With a View to Breaking the Link Between Armed Conflict and the Trade in Rough Diamonds* (Nov. 29, 2001).

and other crime. Lack of transparency in industry operations also facilitates illegal activity. The movement of diamonds from mine to consumer has no set patterns, diamonds can change hands numerous times, and industry participants often operate on the basis of trust, with relatively limited documentation. All of these practices reduce information about diamond transactions. The lack of industry information is exacerbated by poor data reporting at the country level, where import, export, and production statistics often contain glaring inconsistencies.

U.S. control over diamond imports is based on its general control system for most commodities. This control system requires that diamond import documentation include the country of last export—which U.S. import requirements consider the country of origin. Because the current import control system does not require certification from the country of extraction—just from the country of last export—it is not effective in identifying diamonds that might come from conflict sources. Beginning in 1998, rough diamond imports from Angola and Sierra Leone not bearing the official government certificate of origin as well as all rough diamonds from Liberia were banned from the United States.[2] U.S. Customs requires that all shipments from Angola and Sierra Leone have a certificate of origin or other documentation that demonstrates to Customs authorities that the diamonds were legally imported with the approval of the exporting country governments.[3] However, without an effective international system that can trace the original source of rough diamonds, the United States cannot ensure that conflict diamonds do not enter the country.

The Kimberley Process proposal for an international diamond certification scheme lacks some key elements of accountability. We evaluated the scheme using aspects of established criteria for accountability—control environment, risk assessment, control activities, information and communications, and monitoring.[4] While we do not expect the Kimberley proposal to fully address all these elements, this examination provides insights into its ability to deter trade in conflict diamonds. Our assessment of the scheme showed that it incorporates some elements, such as

---

[2] The United Nations Security Council has imposed international sanctions on rough diamond imports from the National Union for the Total Independence of Angola, the Revolutionary United Front in Sierra Leone, and Liberia.

[3] Executive Order 13213 dated May 22, 2001, banned all rough diamond shipments from Liberia for an indefinite period.

[4] The U.S. government, industry, and international entities such as the World Bank accept these internal control standards applied to organizations. See *Standards for Internal Control in the Federal Government*, (GAO/AIMD-00-21.3.1, Nov. 1999), and *Internal Control—Integrated Framework (1985)*, published by the Committee of Sponsoring Organizations of the Treadway Commission and used by the World Bank.

requiring that Kimberley Process Certificates that designate country of origin for unmixed shipments accompany each shipment of rough diamond exports. But some important elements are lacking, and others are listed only as optional or recommended. For example, the scheme is not based on a risk assessment--an essential element. As a result, some activities that would be deemed high-risk by industry experts as well as Kimberley participants, such as the flow of diamonds from the mine or field to the first export, are subject only to "recommended" elements. Additionally, the period after rough diamonds enter a foreign port to a final point of sale will be covered by an industry system in which participation is voluntary and monitoring and enforcement are self-regulated. Other issues relating to accountability are also being discussed by four Kimberley working groups: the establishment of a secretariat; compliance with World Trade Organization rules; sharing of statistics; and monitoring needs. Although the Kimberley Process participants have achieved significant cooperation among industry, nongovernmental organizations, and governments to address trade in conflict diamonds, our work suggests that the participants face considerable challenges in establishing a system that will effectively deter this trade.

BACKGROUND

Conflict diamonds are primarily associated with four countries: Sierra Leone, Liberia, Angola, and the Democratic Republic of the Congo.[5] In all four countries, the production and/or trade of diamonds have played a role in fueling domestic conflict, or, as is the case with Liberia, fueling conflict in neighboring Sierra Leone through the Revolutionary United Front (RUF). Today, Sierra Leone is experiencing relative peace with the aid of the United Nations and other efforts. Nonetheless, diamond mining remains one of the only viable economic opportunities for ex-combatants, and thus experts believe the ability to adequately manage this resource will be important for efforts at establishing long-lasting peace. In Angola, the National Union for the Total Independence of Angola (UNITA) retains control of some diamond production areas, as well as unknown quantities of stockpiled diamonds. And in the Democratic Republic of the Congo, diamonds continue to serve as a

---

[5] Adjacent countries, such as Congo-Brazzaville, Guinea, Cote d'Ivoire, and the Gambia, have all been listed in U.N. reports as countries through which conflict diamonds are smuggled. People named in U.N. reports for their involvement in trading conflict diamonds have been citizens of the Middle East, Europe, and the United States. Also, recent media reports have focused on the possible use of diamonds by terrorists to fund their activities.

source of revenue for armed militias fighting in the north of the country. To date, United Nations sanctions have been targeted solely at rough diamond exports from the RUF in Sierra Leone; Liberia; and UNITA in Angola. Also, both the governments of Sierra Leone and Angola have national diamond certification schemes in which certificates of origin are issued and accompany rough diamonds from their first export to their first import into a foreign country.

Structure of Diamond Industry

The international diamond industry comprises three sectors: mining, rough diamond trading and sorting, and cutting and polishing. This industry structure includes both large and well-organized components as well as small, uncontrolled operations. For example, due to the substantial capital required for deep mining, just four companies mine 76 percent of the world supply of rough diamonds.[6] Yet, across Africa, countless individual diggers mine widely scattered alluvial fields[7] for diamonds. Similarly, while De Beers controls a large percentage of diamond shipments to key trading centers, U.N. data suggest that more than 100 countries worldwide participate in rough diamond exporting. In terms of cutting and polishing, markets have largely evolved to reflect labor costs, with 9 out of 10 rough diamonds cut and polished in India. However, mining countries such as Russia, South Africa, Botswana, and Namibia are trying to expand their cutting and polishing activities to supplement mining revenues.

The Kimberley Process

In May 2000, African diamond producing countries initiated the Kimberley Process in Kimberley, South Africa, to discuss the conflict diamond trade. Participants now include states and countries of the European Union involved in the production, export, and import of rough diamonds; as well as representatives from the diamond industry, notably the World

---

[6] These four companies are De Beers Consolidated Mines Ltd., Alrosa Ltd., Rio Tinto, and BHP Billiton.
[7] Alluvial fields are surface areas containing secondary deposits of weathered volcanic rock called kimberlite deposited by river systems.

Diamond Council,[8] and nongovernmental organizations. The goal is to create and implement an international certification scheme for rough diamonds, based primarily on national certification schemes[9] and internationally agreed minimum standards for the basic requirements of a certificate of origin. The scheme's objectives are to (1) stem the flow of rough diamonds used by rebels to finance armed conflict aimed at overthrowing legitimate governments; and (2) protect the legitimate diamond industry, upon which some countries depend for their economic and social development. U.N. General Assembly Resolution 55/56, adopted on December 1, 2000, requested that countries participating in the Kimberley Process present to the General Assembly a report on progress developing detailed proposals for a simple and workable international certification scheme for rough diamonds.

According to the South Africa Department of Foreign Affairs, the Kimberley Process submitted a report to the U.N. General Assembly in late 2001.[10] The report was accompanied by a proposal for an international certification scheme for rough diamonds dated November 28, 2001, which was to provide the basic elements envisaged for the certification scheme. Participants asked that the certification scheme be established through an international understanding as soon as possible, recognizing the urgency of the situation from a humanitarian and security standpoint. The report also requested an extension of the Kimberley Process mandate to the end of 2002 to enable finalization of the international understanding. Those in a position to issue the Kimberley Process Certificate were to do so immediately. All others were encouraged to do so by June 1, 2002. Further, it was the intention of participants to start full implementation of the scheme by the end of 2002. Finally, a draft resolution seeking an international endorsement of the scheme will be submitted to the U.N. General Assembly for consideration, possibly as soon as late February.

---

[8] The World Diamond Council is an industry association comprising the World Federation of Diamond Bourses and the International Diamond Manufacturers Association, which formed this body expressly to address conflict diamonds.
[9] National certifications schemes have been set up in Angola, Sierra Leone, and Guinea. The High Diamond Council in Antwerp provides technical assistance.
[10] The report has to be translated into the working languages of the United Nations before it can be distributed. This work is almost complete, and the report is expected to be distributed to U.N. members in New York very shortly.

U.S. Participation in the Kimberley Process

In May 2000, the U.S. government established an interagency working group to provide input to and representation at the Kimberley Process meetings. The working group is headed by the Department of State; other participants include the Departments of Commerce, Justice, and Treasury, U.S. Customs Service, Federal Trade Commission, Office of U.S. Trade Representative, U.S. Agency for International Development, National Security Council, Central Intelligence Agency, and the Office of Science and Technology. The United States is currently chairing the Kimberley Process working group on World Trade Organization compliance issues.

NATURE OF DIAMONDS AND NON-TRANSPARENT INDUSTRY OPERATIONS CREATE OPPORTUNITIES FOR ILLICIT TRADE

The illicit diamond trade, including that in conflict diamonds, is facilitated by the nature of diamonds and the lack of transparency in industry operations. Although industry and nongovernmental organizations have made estimates of both the illicit and conflict diamond trades, the criminal nature of the activity precludes determination of the actual extent of the problem. Conflict diamond estimates vary from about 3 to 15 percent of the rough diamond trade and are often based on historical production capacities for rebel-held areas. Some industry experts dispute the larger percentage, believing it includes non-conflict illicit trade.

The Nature of Diamonds Facilitates Illegal Trade

The nature of diamonds makes them attractive to criminal elements. Diamonds are found in remote areas of the world and can be extracted both through capital-intensive deep mining techniques as well as from alluvial sources using rudimentary technology. Individual diggers across west and central Africa mine alluvial fields that are widely scattered and difficult to monitor, a problem made worse by porous borders and corruption. Diamonds are easy to conceal and smuggle across borders, and smuggling routes are well established by those who have done so for decades to evade taxes. Though it may be possible for experts to identify the source of an unmixed parcel of rough diamonds, once diamonds from various sources are mixed, they become virtually untraceable. Identifying the origin of alluvial diamonds is complicated by the fact that the river systems depositing those diamonds run across government- and rebel-held areas as well as national borders. Although rough diamonds can be marked, once they are cut and polished, any form of identification is erased. All of these factors, combined with inadequate customs and policing worldwide, make diamonds attractive to criminal elements who may use them to trade arms, support insurgencies, and plausibly engage in terrorism. Likewise, diamonds can be used as a means of currency in connection with drug deals, money laundering, and other crime or as a store of wealth for those wishing to hide assets outside the banking sector where they can be detected and seized.

Industry's Lack of Transparency also Facilitates Illicit Trade

The flow of diamonds from mine to consumer, referred to as the "diamond pipeline," has no set patterns. Diamonds can change hands numerous times as shown by the fact that the value of world rough diamond exports is three times as large as the value of world rough diamond production. According to industry experts, diamonds are sold back and forth and mixed and re-mixed making tracking a particular shipment through the pipeline and across borders an arduous if not impossible task. Diamonds can be traded in smaller markets and diverted through alternative routes either to disguise origin or in response to low taxes and less burdensome regulations. Thus, the mobility of the trade has also acted as a disincentive for individual governments to implement stricter controls.

Limited transparency in diamond flows is reflected in inconsistent and insufficient data. U.N. data show large discrepancies between export and import data. For example, while Belgium reported selling $355 million worth of rough diamonds to the United States in 2000, the United States reported buying only $192 million worth of rough diamonds from Belgium. U.N. data also suggest that reported world imports of rough diamonds from many countries far exceed those countries' production. For instance, the Central African Republic's production of rough diamonds was worth $72 million in 2000, while global imports from that country totaled $168 million, and the Democratic Republic of the Congo's production was worth $585 million in 2000, while global imports from that country totaled $729 million. Similarly, global imports of rough diamonds from the United Arab Emirates totaled $177 million in 2000, while that country neither mines rough diamonds nor reports having imported rough diamonds from producing countries.

These data inconsistencies can be attributed to a wide variety of factors including:

- differences in how customs officials appraise shipments so that export values differ from import values;
- industry practices such as selling goods on consignment or unloading stockpiles so that trade data differ from production capacities;
- false declarations by importers on where they obtained their shipment, leading to data indicating a country's exports exceed its production; or
- smuggling.

Unfortunately, diamond trade data limitations have been difficult to rectify given that the industry has historically avoided close scrutiny. According to industry experts and government officials, U.S. and international diamond firms do not share trade information freely and business may be conducted on the basis of a handshake, with limited documentation. Furthermore, information problems resulting from industry's lack of transparency are made worse by poor data reporting from many mining and trading nations.

Another factor with the potential to limit transparency in the international diamond industry is the current trend toward merging mining with cutting and polishing activities at the country level. In response to reduced demand and declining rough diamond prices, a number of mining countries are encouraging domestic cutting and polishing. However, when diamonds are cut and polished in mining countries, the source of the rough diamonds used cannot be verified.

THE UNITED STATES CANNOT DETECT CONFLICT DIAMONDS WITH PRESENT IMPORT CONTROLS

Under its current import control system, the United States cannot determine the true origin of diamond imports nor ensure that conflict diamonds do not enter the country. In 1998, the United States began to enhance controls to prevent conflict diamonds from entering the country from U.N. and U.S. sanctioned sources. Since 1998, there have been six diamond-related investigations. However, none of these cases resulted in federal prosecutions relating to diamond smuggling. Without an effective international system to identify the origin of rough diamonds, the United States remains vulnerable to diamonds from conflict sources sent to second countries and then shipped to the United States.

Diamond Imports Subject to General Import Controls; Limited Controls Added to Implement U.N. Sanctions

Diamond imports are subject to the same import controls used for most commodities. Documentation accompanying diamond shipments entering the United States must include a commercial invoice, country of last export, total weight, and value. However, the regulations do not require exporters to specify the country of extraction nor the place of first export. For example, rough diamonds could be mined in one country and traded several times before reaching their final destination. The ability to determine the true source of origin is further impeded because U.S. import shipments can contain diamonds mixed together from numerous countries. Under the current system, Customs would only have documentation citing the last export country.

Until 1998, the United States did not consider conflict diamonds a commodity of focus. But beginning in 1998, the United States put into place import controls to target diamonds documented as originating from the National Union for the Total Independence of Angola, the Revolutionary United Front in Sierra Leone, and Liberia—all of which are subject to U.N. sanctions. Rough diamonds from Liberia have been banned indefinitely from the United States. U.S. Customs requires that all shipments from Angola and Sierra Leone have a certificate of origin or other documentation that demonstrates to U.S. Customs authorities that they were legally imported with the approval of the exporting country governments. However, the controls do not prevent diamonds from these conflict sources from being shipped to a second country and mixed within shipments destined for the United States.

In fiscal year 2000, about $816 million of rough diamonds from 53 countries officially entered the United States through 19 different ports of entry. According to Customs officials, 35 random physical inspections of rough diamond mixed shipments have been performed since 1998. Of these, five cases were found to have minor discrepancies primarily because of incorrect documentation or the diamonds were misdelivered.[11] Customs officials stated that it is virtually impossible to determine the original source of rough diamonds based on physical inspection; thus U.S. Customs officials must rely on the accuracy of the source cited in accompanying import documentation.

CURRENT KIMBERLEY CERTIFICATION SCHEME LACKS KEY ASPECTS OF ACCOUNTABILITY

The Kimberley Process working document describing the essential elements of an international diamond certification scheme[12] does not contain the necessary accountability to provide reasonable assurance that the scheme will be effective in deterring the flow of conflict diamonds. Without effective accountability, the certification scheme may provide the appearance of control while still allowing conflict diamonds to enter the legitimate diamond trade and, as a result, continue to fuel conflict.

The Kimberley scheme primarily provides a description of what participants should do as well as "recommendations" and "options." The document describing the scheme is divided into sections covering definitions, the Kimberley Process certificate, undertakings concerning international trade, internal controls at the national and industry levels, cooperation and transparency, and administrative matters. Elements of internal controls are addressed throughout the document, such as the requirement that the Kimberley Process certificates, designating the country of origin for unmixed parcels, accompany each shipment of rough diamonds and that the certificates be readily accessible for a period of no less than 3 years. However, the scheme lacks key aspects of effective controls, and some "controls" are considered "recommended" or "optional." Some of the areas needing further attention include issues on which agreement has not yet been reached. Working groups have been assigned to address these issues, which include the possible establishment of a secretariat, compliance

---

[11] According to U.S. Customs officials, these inspections were suspended after September 11, 2001, because the agencies' primary focus has shifted to security and anti-terrorism efforts.

[12] *Essential Elements of an International Scheme of Certification for Rough Diamonds, With a View to Breaking the Link Between Armed Conflict and the Trade in Rough Diamonds* (Nov. 29, 2001).

with World Trade Organization rules,[13] sharing of statistics, and the level of monitoring needed.

To assess the current scheme, we looked at evaluations of other international certification schemes and other sources for criteria that can be used to evaluate the Kimberley certification system. We believe the best criteria available are based on standards for internal control that have been developed for organizations.[14] The Kimberley Process participants recognize the importance of internal controls,[15] and the U.S. government, industry, and the international entities such as the World Bank have accepted these standards. While the Kimberley Process is not an organization, the criteria provide useful insights into the ability of the Kimberley Process to achieve basic objectives of accountability and transparency. The guidelines include five control elements—control environment, risk assessment, control activities, information and communications, and monitoring. I will discuss each element and some of the key aspects lacking in the current Kimberley scheme.

Control Environment: A control environment is one with a structure, discipline, and climate conducive to sound controls and conscientious management. The Kimberley scheme faces serious challenges in meeting these criteria.

- Kimberley participants have been unable to agree on the form of administrative support at the international level, whether it is a secretariat or some other mechanism. According to the Kimberley document, institutional arrangements, or the administrative support for the scheme, will be discussed at a future plenary meeting, and no commitments have been made with regard to staffing or funding.[16]

---

[13] Under the Kimberley scheme, participants are to ensure that no shipment of rough diamonds is imported from or exported to a non-participant. However, article XI of the General Agreement on Tariffs and Trade (GATT), 1994, obligates countries to refrain from imposing quantitative restrictions or similar measures on the importation of products from other countries. Two possible exemptions under GATT are being discussed—article XX provides general exemptions and article XXI provides a security exemption.

[14] See *Standards for Internal Control in the Federal Government, (GAO/AIMD-00-21.3.1, Nov. 12, 1999), and Internal Control—Integrated Framework*, published by the Committee of Sponsoring Organizations of the Treadway Commission.

[15] According to the November 2001 Kimberley Ministerial statement, "an internal certification scheme will only be credible if all participants have established effective internal systems of control designed to eliminate the presence of conflict diamonds in the chain of producing, exporting, and importing rough diamonds within their territories...".

[16] Researchers reviewing multilateral environmental agreements have noted that institutional arrangements have come to be seen as crucial to their effectiveness and that the lack of

- Individual participants are required to set up a system of national internal controls and effective enforcement and penalties. It is unclear how and when the capabilities of different participants to do so will be assessed and, where needed, assistance provided. If countries fail to comply with the essential elements of the scheme, then according to the scheme, they can be excluded from trading with participants. However, whether this provision complies with trade agreements such as those under the World Trade Organization has been a point of contention since early in the process and remains under discussion by one of the working groups.

- Political willingness as well as industry commitment to support and implement Kimberley vary. Membership is voluntary, and despite efforts to recruit more members, some key countries have not participated in the Kimberley Process. Further, the United Nations discontinued its "name and shame" policy concerning trade in conflict diamonds because of the lack of clear and consistently applied investigative standards. How the United Nations responds to the Kimberley document and what form the final document will take (an agreement, memorandum of understanding, or some other form) are not known.

Risk Assessment: A risk assessment is a mechanism for properly identifying, analyzing, prioritizing, and managing risks to meet objectives. The Kimberley Process does not include a formal risk assessment and thus participants cannot be assured that appropriate controls are in place. Three potential high-risk areas not adequately addressed in the Kimberley scheme include the following.

- Industry experts and Kimberley participants agree that unless the segment of the diamond pipeline from when the diamond is first discovered in the alluvial field or mine to the point it is first exported is subject to controls, conflict diamonds may enter the legitimate trade. The scheme does little to address this issue, offering only recommendations encouraging participants to license diamond miners and maintain effective security.

---

institutions limits the capacity to monitor states' implementation of and compliance with treaty requirements or to take action when noncompliance is ascertained.

- Industry and others hold stockpiles of diamonds with undocumented sources and the number of diamonds held in stockpiles may be considerable. Since the Kimberley scheme requires information on origin, it is unclear how these diamonds will be addressed. Apparently, any conflict diamond could be claimed as a stockpiled diamond at the scheme's initiation.

- The period after rough diamonds enter a foreign port until their point of sale as rough diamonds, polished diamonds, and jewelry will be covered by an industry system called a chain of warranties in which participation is voluntary and monitoring and enforcement are self-regulated.[17]

Control Activities: Control activities consist of policies, procedures, techniques, and mechanisms that ensure that management directives are being carried out in an effective and efficient manner to achieve control objectives. The Kimberley scheme's inconsistent attention to control activities raises concerns, such as the following.

- While some internal controls are delineated, others are recommended or considered optional without clear justification, and many controls are to be developed at the national level where capabilities and political will differ.

- The industry chain of warranties is based on voluntary participation and self-regulation. Although the scheme requires that all sales invoices of participating industry be inspected by independent auditors to ensure that the diamonds come from non-conflict sources, an audit trail is problematic in an industry where diamonds are sorted and mixed many times.

Information and Communications: An information and communication mechanism is needed for recording and communicating relevant and reliable information to those who need it in a form and time frame that enable them to carry out their internal control responsibilities. Two concerns regarding

---

[17] According to industry officials, the World Diamond Council will strongly recommend that its member organizations require their individual members to make the following statement on all invoices for the sale of rough diamonds, polished diamonds, and jewelry containing diamonds. "The diamonds herein invoiced have been purchased from legitimate sources not involved in funding conflict and in compliance with United Nations resolutions. The seller hereby guarantees that these diamonds are conflict free, based on personal knowledge and/or written guarantees provided by the supplier of these diamonds."

the Kimberley scheme's mechanism for information and communication are as follows.

- Although the Kimberley Process has identified information to be communicated among participants, it has not fully worked out the details of what, how, and when the information will be shared and used. Participants had a great deal of difficulty reaching agreement on sharing statistical data, and a number of issues remain open. The working document states that the content, frequency, timing, format, and methods of handling and exchanging statistical data are to be developed by an ad hoc working group and adopted at a plenary meeting.

- The European Union will function as one trading partner under the Kimberley scheme. It remains unclear how its data will be compiled and shared in a timely manner.

Monitoring: A monitoring mechanism consists of continuous monitoring and evaluation to assess the quality of performance over time in achieving the objectives and ensuring that the findings of audits and other reviews are promptly resolved. Participants had a great deal of difficulty reaching agreement on the need for monitoring. Concerns were raised about sovereignty. A working group is currently addressing this element. The Kimberley scheme's monitoring mechanisms lack details and rely heavily on voluntary participation and self-assessments. For example,

- Monitoring is based on participants' reporting of other participants' transgressions to initiate a verification mission. A participant can inform another participant through the Chair if it believes the laws, regulations, rules, procedures, or practices of that other participant do not ensure the absence of conflict diamonds in the exports of that other participant.

- Review missions are to be conducted with the consent of the participant concerned and can include no more than three representatives of other participant members. Membership and terms of reference of the review missions have not yet been determined. The scheme does not discuss a mechanism for ensuring that the findings of the review missions are promptly resolved.

- No guidelines have been established for developing required self-assessments.
- No system has been proposed for monitoring the industry system of warranties.
- No external audit of the scheme's administration is discussed.

While we do not expect the Kimberley Process proposal to completely address all aspects of accountability, we hope our analysis will be useful in enhancing the scheme's ability to deter the conflict diamond trade. Further, we acknowledge that while the Kimberley Process has brought together industry, nongovernmental organizations, and governments to address a serious humanitarian issue, the participants face significant challenges in deterring the trade in conflict diamonds.

Mr. Chairman and Members of the Committee, that concludes our prepared statement. We will be pleased to answer any questions you may have.

Chapter 6

U.S. INITIATIVES ON "CONFLICT DIAMONDS"*

Office of the Spokesman Director

For over a year the United States has been actively involved in initiatives to curb the powerful and far-reaching impact of the illegitimate diamond trade on African conflicts, particularly in Sierra Leone, Congo, and Angola. "It is time to attack the economy of war that supports illicit arms flows," Secretary Albright said in a special ministerial meeting of the UN Security Council last September. "In many instances, these transactions are fueled by sales of gemstones. . . . Too often the profits fund violence and mayhem--as in Sierra Leone, where illicit diamond profits allowed the Revolutionary United Front to transform itself from a band of 400 to a marauding army of thousands."

The Administration's approach to the complex issue of "conflict diamonds" is to work through a partnership of legitimate diamond-producing states, diamond-consuming and marketing states, and the diamond industry itself. To this end, the Administration has been taking steps to tighten global marketing practices and to build capacity to manage the diamond sector in affected states. At the same time, the U.S. has worked hard to ensure that efforts to address conflict diamonds not harm the stable market democracies

---

* Excerpted from State Department website: www.state.gov

in Africa--particularly Botswana, Namibia and South Africa--which depend heavily on gemstone diamond production.

In recent months, the United States has made headway in engaging the diamond industry, the diamond producing states, and other members of the international community to address this problem.

- On July 5, the UN Security Council--with strong U.S. support--adopted a resolution calling on member states to ban the import of diamonds from Sierra Leone unless those diamonds were exported under a certification scheme to be approved by a Security Council Sanctions Committee. That Committee will hold an open hearing on the role of diamonds in the Sierra Leone conflict July 31-August 1. The United States will speak in strong support of a comprehensive effort to break the link between diamonds and conflict in Sierra Leone and elsewhere in Africa.

- On July 15, U.S. diamond experts--together with UK officials and representatives of the Diamond High Council--traveled to Freetown to work with the government of Sierra Leone on a certification system for diamond exports.

- The U.S. has also taken the lead in establishing Sierra Leone's Commission on the Management of Strategic Resources, committing one million dollars to the effort and providing considerable technical advice.

- On July 19, the World Federation of Diamond Bourses and the International Diamond Manufacturers Association adopted a joint resolution emphasizing the importance of oversight and accountability and proposing a specific program of action to track the flow of rough diamonds. U.S. proposals for a global certification system were instrumental in shaping the resolution.

- In May, at a conference on conflict diamonds in Kimberly, South Africa, African diamond producers, the United States, the United Kingdom, Belgium, and De Beers, among others issue reached agreement in principle on four key points:
    -- the importance of establishing a global certification scheme for diamonds;
    -- the need for a formal code of conduct to govern the practices of the industry, producing states and marketing centers;

-- creation of an independent monitoring agency to supervise implementation of the certification scheme and the code of conduct;
-- establishing a working group to make recommendations on specific mechanisms for implementing these agreements.

The U.S. will continue to work closely with the southern African states to support full implementation of these important recommendations. The Working Group has already met twice, in June in Luanda and in London in July.

- vIn February, De Beers, the international diamond marketing corporation and the world's largest diamond mining operation, announced that it would cease purchasing diamonds from conflict zones in Africa, an important step towards limiting the market for conflict diamonds in Europe, Japan and the United States.

The U.S. has been involved in a series of other activities over the past year to advance prospects for a comprehensive solution:

- Last year the State Department sponsored an international conference in Washington focusing on the economies of war in Angola, Congo, and Sierra Leone, and initiated a direct dialogue with diamond officials from Botswana and Angola;

- The Department sponsored a planning exercise with the Government of Sierra Leone and diamond industry leaders to develop a management plan for the country's diamond resources;

- Together with the UK, the U.S. played a leading role in organizing a meeting in Gaborone, Botswana with diamond authorities, reinforcing consensus support for the twin goals of defining pragmatic measures to combat conflict diamonds while taking special care to do no harm to the legitimate diamond trade;

- The U.S. and the UK have led the push to include the issue of conflict diamonds on the agenda for the G-8 meeting in Okinawa as part of an initiative on conflict prevention;

- The Administration has worked to support the efforts of the UN Experts Panel on UNITA sanctions, under the direction of Canadian Permanent Representative to the UN Robert Fowler.

Chapter 7

SIERRA LEONE: "CONFLICT" DIAMONDS PROGRESS REPORT ON DIAMOND POLICY AND DEVELOPMENT PROGRAM[*]

United States Agency for International Development (USAID) and Office of Transition Initiatives (OTI)

INTRODUCTION

The link between diamonds and armed conflict in Sierra Leone is obvious, and has been exposed, investigated, and deplored by humanitarians, journalists, politicians, and diamond industry leaders. Less obvious are the complex, entrenched relationships between exploitative systems of financial intermediation and resource management, poverty, and the spectacular, mysterious wealth of the diamond trade. Diamonds have facilitated, not caused, armed conflict. Pre-war economic and social injustice, which developed during the war into the illegal, and finally criminal, behavior common of the diamond traffic in Sierra Leone, must be addressed as a complex development problem.[1]

---

[*] Excerpted from United States Agency for International Development (USAID) website: www.usaid.gov/hum_response/oti

[1] This paragraph, and some background text in this report, is drawn from USAID/OTI's Working Paper of 05-08-00, "Diamonds and Armed Conflict in Sierra Leone: Proposal for

In 1999, Sierra Leone's official diamond exports were about $1.2 million, compared to a conservative industry estimate of $70 million as the real commercial value. The other $68.8 million of estimated value was lost to illicit and criminal activity.

Local and international smuggling have enabled the Revolutionary United Front (RUF) rebels and their allies and accomplices to freely and lucratively market diamonds in legitimate international markets. Blood diamonds have found their way onto the sorting tables of mainstream firms in Antwerp, London, Tel Aviv and New York, and presumably into jewelry purchased by customers in the United States, shielded from guilt by the permissive channels of the international market. If smuggling were not such a prominent characteristic of the diamond trade, it would be relatively easy to identify conflict diamonds. UN experts and diamond industry representatives have estimated that conflict diamonds may be as little as 4% of the world trade, in carat weight, whereas smuggled diamonds may account for 20-30%.

Smuggling is a crime against the state: a financial crime of tax evasion. Dealing in conflict diamonds should be considered a crime against humanity. The problems both for the industry and for governments are considerable. Going after a small number of criminals rash and despicable enough to deal in conflict stones is exceedingly difficult, when they must be sought among the vast numbers of legitimate traders on the world market. The diamond industry has been categorical to declare it will banish permanently from doing business in any bourse, worldwide, any industry member found trading in conflict diamonds. But the trade goes on, and no cases have been discovered.

GENERAL BACKGROUND

In December 1999, USAID's Office of Transition Initiatives (OTI) began providing technical assistance to the Government of Sierra Leone (GOSL) to develop new diamond policies and establish new mining and exporting operations that would address the link between diamonds and the war. In March 2000, USAID/OTI, with consultants from Management Systems International (MSI), facilitated a strategic planning workshop in Freetown, attended by GOSL cabinet members and technical personnel

---

Implementation of a New Diamond Policy and Operations", posted on www.usaid.gov/hum_response/oti.

relevant to the diamond sector, representatives of civil society and the rebel Revolutionary United Front (RUF), and international diamond industry leaders and experts, including De Beers. Following this exercise, De Beers led the industry in May with its proposals on how to try to identify conflict diamonds and prevent their exploitation by rebels in Africa, and particularly in Sierra Leone. USAID/OTI presented a working paper on May 8, 2000 that summarized a way forward, which was subsequently published by the GOSL under the title "Guidelines for the mining and export of diamonds in Sierra Leone", and adopted by the GOSL as its policy framework. It has been systematically implemented since mid 2000.*

The international momentum on "conflict" diamonds increased, with a number of multi-sectoral and multi-national technical meetings. At a technical meeting in Kimberley, South Africa, in mid May, representatives of USAID and the Diamond High Council (HRD) of Belgium discussed the development of a Certificate of Origin for Sierra Leone, to be modeled on the Angolan Certificate which the HRD had developed with the Government of Angola. Proposed by the United Kingdom (UK), United Nations Security Council Resolution 1306 (2000) banned the import of all diamonds from Sierra Leone as of July 5, 2000, until an effective certification system was fully operational.

By this time, the US Government was working in a trilateral approach, with Belgium and the UK, to assist the GOSL with new export policies to control "conflict" diamonds, within the framework of the "Guidelines" paper. A trilateral (US, UK, Belgian) technical mission in Freetown on July 14-15, 2000, served to unite assistance efforts on behalf of the GOSL. Subsequent technical meetings in Antwerp (July 18), Washington (July 27), and New York (July 28) brought the Certification design for Sierra Leone to closure.

On July 31 and August 1, 2000, at informal UN hearings in New York, the GOSL presented the basic elements of a new mining and export regime, which it hoped would qualify for an exemption to UN Resolution 1306 (2000). In mid October, the major elements of the new export regime were operational, and the Security Council granted the GOSL an exemption to Resolution 1306, to permit exports to resume. There was widespread acknowledgement that the new regime would require continuing international assistance, effort by the GOSL, and complementary actions by

---

* The "Guidelines" document is the same text as USAID/OTI's Working Paper of 05-08-00, op. cit.

the international diamond industry, in order to effectively control "conflict" diamonds.

The World Diamond Congress, meeting in Antwerp on July 17-18, 2000, was opened with a presentation by Congressman Tony Hall on "conflict" diamonds. Led by the International Diamond Manufacturer's Association (IDMA), the Diamond Congress condemned the marketing of "conflict" diamonds, and laid the foundation to constitute the World Diamond Council (WDC), whose inaugural meeting was held in Tel Aviv on September 7-8, 2000.

International human rights activism has been important in bringing the plight of abused civilians in Sierra Leone to the attention of international political leaders, thus creating the conditions for diplomatic and peace-keeping support to match what was already being done by the international community on humanitarian response. An important contribution to peace by human rights activists is their work to expose the link between armed conflict and diamonds in Africa, pioneered by the London-based nongovernmental organization (NGO), Global Witness, and the research done by Partnership Africa Canada (PAC). There is robust engagement of US-based and international NGOs in this cause, including World Vision, Physicians for Human Rights, and Amnesty International.

SMUGGLING AND CONFLICT DIAMONDS

Conflict diamonds are a short-term problem for Sierra Leone, which will end with the war. Smuggling, exploitative systems of production and marketing, and undeveloped financial services have caused the depravation in diamond-producing areas that has facilitated the use of diamonds for war. USAID's approach has been to support policy change that addresses these long-term problems, while also providing a solution to conflict diamonds. When the RUF no longer control Tongo Field and Kono, there will no longer be conflict diamonds from Sierra Leone. But if there is still rampant smuggling, Sierra Leone will not have addressed the fundamental causes of the war, and peace will not bring a chance for development. This justifies USAID's broad approach.

The GOSL has established a "clean" channel for diamond exports, and has adopted a series of policies aimed at reducing the incentives for, and increasing the costs to, smuggling. Law enforcement capabilities against smuggling in 2000 were too weak to make an enforcement effort a realistic approach to smuggling. After the war, as the country's security structures improve, curtailing smuggling through law enforcement may become feasible. The imposition of UN sanctions, and industry announcements against conflict diamonds, such as those of the World Diamond Council, have helped by potentially increasing the costs of dealing in conflict diamonds.

USAID/OTI's approach, through public policy reform, is to reduce the incentives for smuggling, and gradually transform the diamond sector into one in which acceptable standards of good business practice prevail. USAID has similar interventions in a number of countries to help governments address the problem of corruption. Successful anti-corruption programs take an incremental approach, and deal with reforming the systems or practices that enable or permit corruption. The monitoring initiatives that are briefly described at the end of this report are part of the process of reform, and also constitute a community effort that will contribute to law enforcement.

In September 2000, USAID/OTI produced a summary Action Plan. Table 1 describes the desired "outputs" of this assistance effort.

What follows in this chapter is a description of progress to date, including the categories listed below:

- Licensing;
- Export documentation, procedures for a "clean channel", and identifying conflict stones at the point of export;
- Foreign exchange and banking policy;
- Export Performance; and
- Monitoring and detecting smuggling.

Table 1. Summary Of Outputs
Usaid Assistance For Diamond Policy And Development

Policy Outputs	Process Outputs	Peace & Development Outputs
Export procedures: Certification of Origin Valuation procedures Product, document, and information handling	**Transparency:** Eliminate non-disclosure and secrecy in income, foreign exchange, permits, and all kinds of transactions	**Monitoring of "conflict" diamonds:** exporters must have documentation to trace their rough diamond purchases, indicating the source is "non-conflict"
Licensing: Mining Export & buying agents	**Accountability:** - Public servants and trad. Leaders accountable to GOSL and communities - Private sector must report activities and earnings to GOSL fiscal & mining authorities	**Information/Training:** program of information to producing communities about control of "conflict" diamonds; confidence-building that they have the power to approve "conflict-free" transactions; training of diggers & communities in valuation
Regulatory framework for fair practices and good performance: Public auctions to prevent price/valuation collusion Licensing and banking	**Anti-Corruption:** constant vigilance to anticipate weak links in systems and detect and denounce possible corruption for investigation	**Income Distribution:** establish distribution of % of fiscal receipts to producing communities to create their stake in correct disclosure & valuation
Banking: Commercial transactions Foreign exchange Credit policy	**Checks & Balances:** Monitoring for transparency, accountability, and anti-corruption Cross-checks: - Domestic/international - Trad. Soc./NGO/GOSL	**Credit Program:** enable diggers/mining groups to obtain credit for working capital through banks
Fiscal: Taxes & fee collection Tax distribution		**Participation:** org. of communities for approving legitimate transactions & role in fiscal decisions; inclusion of all stakeholders
Law enforcement/ justice: Corruption control Crime detection & prevention		

BACKGROUND ON DIAMOND MINING AND MARKETING IN SIERRA LEONE

Sierra Leone's diamond production is alluvial, and is "mined" over a vast area of the country's territory. See the attached map. Prior to the coup of May 1997, there were three diamond marketing centers: Bo, Kenema, and Koidu, the center of diamond-rich Kono District. Koidu was destroyed by the war in 1998, leaving Bo and Kenema as population and marketing centers. These are also the provincial capitals of the Southern and Eastern provinces, respectively. Both were safe havens throughout most of the conflict, but since the May 1997 coup, Kenema has been a battleground between the RUF and Civil Defense Forces (CDFs), militias loyal to the Kabbah Government, with Kailahun and Kono associated with tight RUF territorial domination. Bo is the heartland of the Mende Kamajors, the basis of the CDFs, that swept the RUF out of the South and parts of the East in 1996, and have maintained control ever since. Thus, the RUF have been in control of mining areas in most of the Eastern province and Kono since 1997; the CDF have controlled mining areas in the South since 1996. The only areas of historical importance to mining that are currently controlled by the RUF are Tongo Field and the Kono area. These areas are where the largest stones come from, being closer to the kimberlite source of Sierra Leone's alluvial production. After the May 1997 coup, the RUF also gained access to Makeni, where there are some new diamond-producing areas, not of historical importance.

Alluvial diamonds are found by "diggers", who manually, or with rudimentary equipment, sift through soil and sand, digging holes up to 30 feet in depth, in areas where they think it is most likely to find stones. Only men are diggers; women are farmers and petty traders, and service the household. Most diggers are the poorest of the poor, doing body-breaking work with no certainty of finding any stones, but with the illusion of uncovering a large stone that will provide wealth for life; not a common result for diggers.

Land is communally owned in Sierra Leone. "Leases" are managed and rents collected by traditional paramount chiefs. Diggers obtain permission to dig in specified areas from the respective chief. Because of the requirement to pay a land use fee, the diggers are generally financed by "dealers". Dealers are business people who manage groups of diggers by advancing them food, tools, and basic household goods, which they deduct from the proceeds of sales of the stones the diggers turn over to them. Over time, poverty has conspired with ignorance to create a system of virtual servitude. A new observer to the scene can hardly imagine how such exploitation can still exist in the 21st century.

The "dealers" sell to "exporters" and "agents" of the exporters, who buy and export the stones. Stones can change hands several times among dealers before they are finally exported. Prior to the coup of May 1997, some buyers declared their stones to the Government Gold and Diamond Office (GGDO), which valued them for the purpose of export taxes and statistics. Others smuggled the stones out of the country, with no documentation or registration, and were able to market them, eventually, in mainstream international markets.

During the war, the roles of dealers and buyers became quite murky. Buyers formerly would have been licensed firms, declaring at least some of their transactions to the government, obtaining export documentation, and dealing in the official foreign exchange market. Throughout the early part of the war, and up until the time of the 1997 coup, Bo, Kenema, and Koidu were safe havens, and Freetown and Makeni were scarcely even affected by the war. The smuggling of diamonds co-existed with the presence of mainstream, "legitimate" Lebanese merchants, whose (probably not fully disclosed) transactions were not part of the war economy of the RUF, but were just part of an inefficient system that evolved to accommodate the many petty stakeholders in corruption.

The final plunge into the obliteration of mainstream commerce occurred when most of the Lebanese merchants fled. What remained was a small cadre of diamond runners who now serviced only the war-makers. Those doing business during the coup had no pretext or cover of doing legitimate business, and even in 1998, after the restoration of the Kabbah government, this lapse from a shadowy, mixed economy into near total crime continued, as evidenced by Sierra Leone's diamond export statistics of 1999. See the data presented in the section on export performance.

Even before the war started in 1991, the diamond market in Sierra Leone followed a downward spiral of degradation of legality (evidenced by declared value of exports) for many years. Two interesting summaries of this history have been recounted by the political scientist and historian Will Reno and by the authors of "The Heart of the Matter".[*]

The return of most of the dealers and exporters who did business before the coup was gradual. The volume of official exports in 1999 was tiny, but a

---

[*] William Reno, Department of Political Science, northwestern University, Evanston, Illinois, USA; "The Heart of the Matter", Ian Smillie, Lansana Gberie, and Ralph Hazleton, Partnership Africa Canada (PAC), Ottawa, Canada, January 2000. E-mail: *pac@web.net*.

positive number. In 2000, official exports started to trickle upwards, and Bo and Kenema seemed to be doing business as usual, such as the old days; except for the fact that Kono District, Tongo Field, and other less important areas in the Eastern and Northern Provinces, were occupied by the RUF.

LICENSING*

New GOSL Policy on Licensing

The need for reform of the diamond sector, to address conflict diamonds, created the need to review the structure of the market in Sierra Leone, including the roles of business people associated with diamonds, and their agents.

One option considered last year by the Government of Sierra Leone was to license only one, or a very few, world-class firms to mine and export diamonds. This would have transferred the responsibility for controlling smuggling and conflict diamonds from the GOSL to one (or a few) private, reputable firms. At first glance a simple and thus appealing solution, it was deemed impossible to achieve, given the alluvial nature of diamonds and rebel control over a large part of the diamond-producing areas. Implementation would have required massive enforcement by foreign security forces. The GOSL reported that there were no takers for this option.

Mining Licenses

In the GOSL's presentation to the United Nations, at informal hearings of the Sanctions Committee on July 31, 2000, the Minister of Mineral Resources stated that mining would be licensed only in areas under GOSL control. Approval of mining licenses would follow the historical practices, requiring persons to first obtain land-use permission from local traditional authorities, and then to issue licenses to persons who met the minimum requirements, with the Ministry acting more as a registration than as a regulatory body. This system would enable business people who had historically been exporting diamonds to continue to do business. A copy of

---

* Part of this section is taken from pages 4 and 5 of the USAID/OTI Working Paper of 05-08-00, previously cited.

the GOSL policy on licensing is attached, as is a fee schedule for all categories of licenses and fees associated with mining.

Representatives of civil society organizations have stated that the fee structure presents an entry barrier for most Sierra Leoneans, who do not have enough savings to pay for mining or dealer's licenses. They are prey to higher-income business people who provide financing to them in kind, creating a relationship something like indentured servitude. The issue, as expressed by civil society representatives, is that low-income people, who cannot afford a license, feel they do not have the opportunity to circumvent dealers who, in their view, have historically exploited them. The presence of some new international buyers up country, alternative sources of credit, and better information on how to value rough stones could help prevent low-income Sierra Leonean diggers and miners from being exploited by dealers.

Export Licenses

In December 2000, the Ministry of Mineral Resources adopted a new policy for export licensing, authorizing eight "umbrella" export licenses, with a fee of $50,000 per year for each one. Under the new policy, each license holder can have up to five agents, who must also be registered. Further, the Ministry has allowed licenses to be shared, or subdivided. The model text for an exporter's agreement with the GOSL was published in the press on December 7, 2000, and is attached. As of March 30, 2000, the status of signed export license agreements was as follows:

United States: 1 license holder, 2 co-holders, each have paid $25,000, 1 license issued
Belgium: 1 license, 3 co-holders, each paid up, 2 licenses issued
South Africa: 1 license, 1 holder paid $25,000, 1 license issued
India: no one paid
Israel: 2 holders, each paid $25,000, 2 licenses issued
Sierra Leone: (citizens): 2 licenses: 1 corporation (NAMINCO) and 17 individuals joined together for the other
Sierra Leone: (non-indigenes): 5 individuals, each paid $10,000

The changes in the system have created some confusion about who is authorized to transact business in diamonds. The intention of the Ministry at first seems to have been to limit the number of license holders. This intention has not been accomplished due to the multiple sub-divisions of the eight export license groups. Given the weak enforcement mechanisms, limiting license holders probably would not have reduced smuggling. Instead, limiting access could result in corruption, by making license holders "gatekeepers" to diamond transactions. A first step to reducing smuggling is probably to get as many of the diamond players to register, and do business through official channels.

With one transitory exception some years ago, NAMINCO is the only case in Sierra Leone in which an indigenous business group has put together enough financial resources and business experience to seriously pursue mining and exporting. Joint ventures may be one of their future options, however for the moment they seem to have pooled enough equity to constitute a viable business without foreign partners. Another outcome of this process of reform is the licensing of U.S.-based entities in the Sierra Leonean rough diamond sector, who have declared their intention to develop ways to improve the income of small-scale miners, and to provide an example of scrupulous adherence to the ban on conflict diamonds.

The major deficiency of the export license scheme at present seems to be that the policy has not been clearly defined and consistently practiced. Modifications have been improvised, creating inequities in the fees paid. The Ministry declared that license renewal would be contingent on export performance, setting a target of $10 million of exports per year for each of the "umbrella" groups. Specific performance targets need to be specified for every license holder, and a quarterly review should be done so that when the annual review comes up, there will be no surprises.

The GOSL has conducted meetings with exporters and dealers, in Freetown and in Bo and Kenema. However, a manual or handbook of information for exporters, dealers, and miners is recommended, as well as further meetings, to ensure that there is uniformity of treatment and standard information supplied to all parties.

CERTIFICATION OF ORIGIN

As reported above, United Nations Resolution 1306 (2000) prohibited the direct or indirect import of all rough diamonds from Sierra Leone, until such time as the Government of Sierra Leone put into operation an effective Certificate of Origin regime for trade in diamonds. Using a trilateral approach, the governments of Belgium, the United States, and the United Kingdom, and the Diamond High Council (HRD) of Belgium, provided assistance to the GOSL to design, and help put into use, new export documentation and procedures. This work was well advanced before the UN adopted sanctions, therefore qualification for an exemption took only three months.

The centerpiece of the new export procedures is the Certificate of Origin. It was modeled on the Angolan Certificate of Origin, developed for the Government of Angola with technical assistance from the HRD. The trilateral team introduced significant innovations for electronically tracking exports and imports of parcels of diamonds, and for better identification using digital photographs. The GOSL presented the proposed new system to the UN at informal hearings on July 31 and August 1, 2000. An exemption under paragraph five of UN Resolution 1306 (2000) was granted in early October 2000 to permit the resumption of exports from Sierra Leone, if accompanied by the new Certificate of Origin issued by the GOSL.

As of October 2000, when Sierra Leone was granted an exemption to UN Resolution 1306 (2000), all diamond exports must be accompanied by a Certificate of Origin. In order to obtain a Certificate of Origin for legal export, the diamonds must be legally mined. Legally mined means they come only from areas under GOSL control, and are the product of a chain of legally authorized transactions, from use of the land, permission to mine, purchase by authorized dealers and agents, and export by licensed exporters. This chain does not differ substantially from the policies in effect before sanctions, however there is more scrutiny on compliance with GOSL regulations, and more still is expected as new monitoring practices are developed.

Documentation

The Certification of Origin regime consists of documentation on security paper, with two numbered, detachable slips. One of the numbered slips is inserted in the parcel of goods, and the other is a receipt confirmation, that should be physically returned by the importing authority to the exporting authority in Sierra Leone.

The Certificate has four signatures, as follows:

- Government Gold and Diamond Office (GGDO): the GGDO performs the official valuation of exported rough diamonds, and is responsible for recording and filing all information, and keeping records of the Certificates of Origin. The GGDO valuator is an expert capable of identifying the origin of stones, from the various diamond-producing areas within Sierra Leone.

- Minister of Mineral Resources: his signatures attests that the exporter is duly licensed; any investigation or monitoring of the chain of transactions should be done by Mines Ministry personnel, to document that all transactions from mining to exporting were conducted by duly licensed parties.

- Governor of the Central Bank: his signature attests to the correct registration of foreign exchange, and enables tracking of the diamond exporter's "account", matching the value of diamonds exported and the value of foreign exchange imported.

- Customs Official: represents the fiscal authorities, attesting that export taxes and fees have been collected.

Electronic Tracking and Data Base

An electronic tracking system links the exporting authority (in Sierra Leone) to the importing authority, and provides advance information, sent electronically, on each parcel and Certificate of Origin. The HRD designed and installed the electronic tracking system, with a low-level security encryption incorporated in a basic office system of computer software. The HRD supplied the hardware and software to link Freetown and Antwerp, and pledged to assist other diamond importing countries to link up to the "network", at the expense of the importing country. Israel has requested information to do so, but is not yet linked.

The Certification system provides that electronic confirmation by the importing authority be made to Freetown, as well as documentary confirmation via the return of the confirmation slip, which is a detachable part of the Certificate of Origin. The physical documentation, electronic tracking and digital photographs are managed by the GOSL Government Gold and Diamond Office (GGDO). Unused Certificates of Origin are stored in the vault of the Bank of Sierra Leone (Central Bank). The offices of the GGDO are in the Bank building.

This system provides for export and import statistics on diamonds of Sierra Leone origin to be readily available.

A strict interpretation of UN Resolution 1306 (2000) and the exemption granted for GOSL exports would provide that only importing authorities of countries that are willing to adopt the Certification system, and be linked into an integrated Data Base, can receive exports from Sierra Leone without violating UN Resolution 1306 (2000). That would imply that diamonds can only be consigned at the present time to an importer in Belgium, as only Antwerp is linked to Freetown.

A loose interpretation, would require Sierra Leonean authorities to issue Certificates of Origin, but would not require the data to be confirmed by the importing country. Therefore, any country could import diamonds from Sierra Leone, as long as the parcel was accompanied by a Certificate of Origin issued by the GOSL.

Since October, there have been several cases of exports duly documented, consigned on the Certificate of Origin to London, Israel, and the United States. This matter should be clarified with the GOSL and with the UN Sanctions Committee. Given the nature of the system as a start-up operation, it should not be interpreted as a deliberate violation of sanctions for this to have occurred.

If the electronic tracking system is to be meaningful in collecting statistics, no transactions should be done outside of a linked system. This would mean that new countries that wish to join the system would have to agree to become part of the electronic loop, agree to acknowledge and exchange the information required by the system, and agree to return the confirmation slips to the GGDO in Freetown.

There were some start-up problems to the electronic tracking and documentation systems between Freetown and Antwerp, such as lack of confirmation of some parcels, return slips not being sent, and inadequate filing systems in Freetown. Most of these problems were addressed during the first three months of the systems' operations. Periodic evaluations should be done.

The electronic tracking system is a system to exchange and confirm the information generated by the GGDO at the source. At the present time, there is no routine or obligatory examination or valuation done of the parcels by the importing authority in Antwerp.

Digital Photographs

Digital photographs are part of the electronic data system, and provide a visual description of the parcel to complement the written description by class, weight, and value, as recorded in the Certificate. Also, digital photographs have been useful to provide close-up photographs of diamonds suspected by the GGDO of coming from "conflict" zones. The photographs could be forwarded to a panel of "conflict" diamond experts for their opinion, were such a panel created. This process of consultations of opinion has been successfully done on an informal basis on at least one occasion since October. The quality of the photograph was deemed sufficient for experts to render an opinion.

Use of Certificates of Origin

Certificate of Origin number 0001 is dated October 23, 2000. The use of Certificates of Origin, since the system began, is as follows:

Table 2. Use of Certificates of Origin: Sierra Leone

Month	# of C.O.'s	C.O. Numbers	Carat Weight	Value in US$
Oct '00*	17	000001-000017	28,450.60	$ 4,470,424.61
Nov '00	6	000018-000024	12,128.75	$ 1,079,695.58
Dec '00**	8	000025-000032	9,702.16	$ 983,014.60
Jan '01	7	000033-000039	13,486.10	$ 1,991,773.84
Feb '01	14	000040-000053	15,384.67	$1,909,276.29
Mar '01	11	000054-000065***	20,055.63	$2,685,334.87
Total			99,207.91	$13,119,519.59

* Includes export of the "stockpile": diamonds held by exporters between July-Oct, when UN sanctions were in force, and exported under the exemption granted by the UN Sanctions Committee
** Certificates 000026 and 000030 had not been used as of March 1, 2001, but the numbers have been allocated to specific parcels of goods
***Certificate 000060 was cancelled due to an error being recorded on it

As important as the adoption of the new certification scheme is, it forms only part of a larger effort to control "conflict" diamonds, other elements of which are described below.

GGDO VALUATION

The identification of "conflict" diamonds at the source (point of export) is now being done in Sierra Leone for the "clean" channel of official exports. Every single diamond is inspected both by the GOSL expert valuator of the GGDO, and by an independent, expatriate valuator, also expert in identifying the source of stones, as described below.

Additional experts, to inspect diamonds at the point of import, could be useful for the following functions:

- As an audit, or quality control practice, which could be accomplished by random inspections of parcels;
- To provide additional opinions, if there is a difference of opinion on the identity of a stone between the GGDO and the independent valuator, or between these two valuators and the owner of the merchandise, in the case of a confiscation.

The Government Gold and Diamond Office is responsible for inspecting and valuing all goods. The exporter presents his parcel of diamonds with the "Schedule B" form, which contains the carat weight and his estimated value per carat of sorted goods. The categories for sorting are specified by the GGDO. A description of the procedure and sorting categories is attached.

The GGDO valuator inspects the parcel in the exporter's presence, verifies weights, and reviews the goods to be certain they have been correctly sorted. He attaches his own estimation of value per carat to each category presented. If there is a discrepancy in the value of the parcel, the GGDO uses the higher of the estimates for the purposes of levying export taxes.

The GGDO's description of the parcel is noted next to the exporter's description on Schedule B, which is attached to the Certificate of Origin. The summary of the parcel's description is recorded on the Certificate of Origin.

Once the valuation has been completed, the parcel is photographed as sorted, the goods are packed in a box supplied by the GGDO, sealed, and returned to the exporter for shipment.

INDEPENDENT VALUATION

Independent valuation is perhaps the single most important addition that the GOSL has made to the exporting process, as an aid to building confidence in the integrity of the valuation process, and for the identification of "conflict" stones. The independent valuator is a full-time expatriate diamond expert who independently values every single parcel, and looks at every single diamond being officially exported with GOSL documentation.

The independent valuator is the Belgian firm of Zurel Brothers, whose principle is Bryan M. Zurel. Mr. Michel Lempel of Zurel Bros. is in residence in Freetown, and began his duties the week of February 5, 2001, at the GGDO. He is present during the valuation done by GGDO, but performs his valuation separately, not making comments during the GGDO valuation. Mr. Lempel lived in Sierra Leone for several years about 10 years ago, and is well acquainted with diamonds from all parts of Sierra Leone. He has approximately 40 years of experience as a diamond buyer and valuator.

The selection of the independent valuator was done as follows. The GOSL asked the Diamond High Council of Antwerp for recommendations of competent, responsible firms. Two firms were recommended. The GOSL asked these firms for bids, and chose the firm with the lower bid. An open, advertised process to receive bids was not done. A three-year contract, with an annual review was signed by the GOSL. Payment for the services of the independent valuator will be made from the proceeds of diamond export fees, for which purpose the GOSL has earmarked 0.5% of the export value of diamonds. The total export taxes are 3%.

It is important to acknowledge that it was the Minister of Mineral Resources that requested, in July of 2000, that the "Trilateral Technical Team" facilitate the services of an independent valuator. The HRD assisted by recommending firms. It was the GOSL that designated .5% of the export taxes to pay for these services, and committed to a three-year contract with the independent valuator. This indicates the intention to address possible corruption within Sierra Leone, as it creates an independent check on both valuation and identification of conflict stones at the point of export.

IDENTIFICATION OF "CONFLICT" STONES AND CONFISCATION

The official GGDO valuator is responsible for identifying stones he believes originate from areas under rebel control. He has the expert knowledge of Sierra Leonean diamonds to do this, however this is a subjective judgement. There is no scientific test available at the present time to definitively determine the exact geographic origin of a diamond. The independent valuator also has the expert knowledge to identify the probable source of a Sierra Leonean diamond. He can validate a judgement of the GGDO valuator, or could make an identification not detected by the GGDO. The GOSL should confiscate "conflict" stones.

Smuggled diamonds are also subject to confiscation. Even before the issue of "conflict" diamonds became a public policy concern, Sierra Leone had rules and regulations that apply to confiscated goods. Confiscated stones, whether they are smuggled diamonds that originate from areas under GOSL control or "conflict" diamonds, are sold at auction, and 40% of the proceeds goes to the person responsible for the confiscation; 60% goes to the GOSL treasury.

To date, the GGDO has not confiscated any conflict stones. Since October 2000, the GGDO has twice made an identification of stones that are suspected to be from Tongo Field, an area under rebel control. At the time of these two cases of tentative identification, the independent valuator had not yet been retained. The GGDO valuator reports he was not certain of his own judgement. Therefore, the stones were individually wrapped and photographed, and the GGDO sought an opinion of the Diamond High Council in Antwerp. In one case, further experts were consulted by e-mailing the digital photograph. The exporter was advised that the GGDO had a suspicion about the origin of these stones, but they were not confiscated. An investigation of the exporter's buying records was not done.

The GGDO valuator reported the week of February 5, 2001, that he has not seen any further questionable stones, after the two cases cited above. He suggested that licensed exporters are either not buying such stones, or are not presenting them to the GGDO for Certification. Therefore, at the present time, according to information from the GGDO, any diamonds from rebel-controlled areas that are leaving Sierra Leone are being smuggled. They are not entering the GOSL "clean" channel, as valued and documented by the Sierra Leonean Certificate of Origin.

FOREIGN EXCHANGE AND BANKING POLICY

Foreign Exchange

Exchange rate distortions were cited in 1999 by international diamond firms as an important historical condition driving legitimate players away from Sierra Leone. This has been corrected, and diamond transactions can now be conducted officially in US dollars in Sierra Leone.

The Bank of Sierra Leone (Central Bank) has achieved stabilization of the exchange rate, and significant currency revaluation has taken place in the last six months (from about Le. 2,500 ["Leones"] to the dollar to about Le.1,750). A policy of weekly auctions of foreign exchange has contributed to this.

Another element of new policy involves ensuring that foreign exchange circulates in the Sierra Leonean economy, creating a multiplier effect. The Central Bank, as one of the signatories to the Certificates of Origin, is responsible for verifying that foreign exchange is brought into the country, obtained through reputable international banks, in approximately the same value as diamond exports, minus fees. The BSL requires an exporter to register imported dollars with a commercial bank, and to provide the BSL with a coded telex confirmation indicating the source and amount, if it is in cash dollars. A summary of the Bank of Sierra Leone's policy on foreign exchange and banking for diamonds is attached.

During November and December of 2000, there were considerable start-up problems between the Bank of Sierra Leone, the Ministry of Mineral Resources, and the GGDO. The Bank refused to sign several Certificates of Origin until the exporter complied with its rules. Exporters argued that they were not informed of the new rules in advance, and that this was causing costly delays to their exports. A normal flow of documentation seems to have been established since then, and the GGDO reported during the week of February 5, 2001, that exporters were now presenting their documentation with full compliance.

Income and its Multiplier

The problem of the low multiplier in the Sierra Leonean economy is a way of saying poverty is constantly reproducing itself despite the presence of spectacular wealth. The low multiplier relates to the amount of income staying in Sierra Leone, and to the way that income is or is not being utilized for investment. Getting more money into the hands of Sierra Leoneans is part of the solution. Too much of the value of the diamond trade belongs to people who are not reinvesting in Sierra Leone. Another part is to have more efficient financial intermediation: a term that means getting money from the pockets of those who have money to those who use it for production; not by confiscation, but by lending and borrowing. This is standard banking practice, but it has been lacking in Sierra Leone.

More important than tax collection, the most important benefit of legitimate diamond operations to the Sierra Leonean economy is the income this should generate. If the estimated $70 million per year of Sierra Leone's diamond value were paid to diggers and dealers, and deposited into banks in Sierra Leone to be invested and consumed in Sierra Leone, this amount of income and its multiplier would constitute an important post-conflict resource. The measures taken by the Bank of Sierra Leone to require exporters to demonstrate foreign exchange inflow are a first step to ensure that the export value of diamonds in the official, "clean" channel circulates as income in Sierra Leone.

Some training of potential borrowers in how to use commercial banking would be useful (literacy is not required, just knowing the basic rules), as well as mechanisms to pool the risk of diggers. Diamond digging, in the aggregate is not risky; but each individual's activity for a finite period of time is risky. Either risk is pooled by the credit provider (by building in a margin for individual default) or it is pooled by the user. Pooling risk by the user seems to be a more attractive option, as this method provides for group or peer compliance with disclosure. This is also consistent with the extended family and community safety-net concept of the traditional society.

Commercial banking needs a lot of work in Sierra Leone. However, the supply side of commercial banking is gradually responding to the demand of the new diamond players, particularly the U.S. based exporters. The development of commercial banking has no doubt been affected by the high-risk environment of a decade of war, including pillaging in Freetown, Bo, and Kenema, and the total destruction of Koidu.

Commercial banks are still not providing a full range of banking services at reasonable rates in Freetown or up country locations. It is important for world-class, competitive banking to be available in order to attract and keep world-class diamond buyers. Reliable safe deposit and checking account services should be available up country as well as in Freetown, and might come into use by some of the new exporters. If such services were available, the Bank of Sierra Leone could further tighten up its requirements for handling of cash. For example, it could follow the common banking practice of the U.S. and Europe, and prohibit large quantities of cash from being carried by individuals without proper documentation. This would enable investigation into smuggling, and would help to differentiate legitimate diamond traders from smugglers.

PENALTIES AND ENFORCEMENT

More work needs to be done in Sierra Leone on law enforcement. The assistance to the GOSL provided by USAID has not covered this area to date.

The GOSL may request technical expertise on money laundering and international financial crimes to assist in the detection of smuggling, money laundering and other financial crimes.

EXPORT PERFORMANCE

Export data is available, but it is not easy to interpret, as there are so many atypical factors that have been present during 1999 and 2000. Historical data from the 90's cannot easily be used to establish production and marketing cycles for the same reasons.

In early 1999 there was no cease-fire, and violence in Freetown in December '98 and January '99 caused many business people to evacuate, particularly expatriates. Whereas Bo has been in GOSL control since about 1996, the war has waged all around it except to the South.

Some dates and events of impact on business and mining conditions are presented as follows:

May '99	Cease-Fire between the GOSL and the RUF
July '99	Lome Peace Agreement between the GOSL and the RUF
Oct '99	Approval by the UN of a Peace-Keeping Operation (UN-PKO)
Dec '99	Arrival in Freetown of first UN peace-keepers
Jan '00	Deployment of UN-PKO up country, and first seizure of UN weapons by RUF
May '00	About 500 UN troops taken hostage by RUF
May '00	Disarmament and Demobilization camps empty out as former army and rebels re-arm
June '00	British show of military force in Freetown and off-shore
July '00	British military unit taken hostage by "West Side Boys" (faction of recalcitrant SLA/RUF/urban thugs)
Aug '00	West Side Boys defeated by British operation (diamonds confiscated)
Aug '00	RUF announce new interim leader and declare they will open roads to GOSL and UN deployment
Sept '00	Cross-border violence from Liberia into Guinea at Macenta – refugees affected
Nov '00	Abuja Cease-Fire Agreement signed between RUF and GOSL
Dec '00	Cross-border violence from Sierra Leone into Gueckedou and other parts of Southeast Guinea, near diamond producing areas of Guinea – refugee crisis deepens
Dec '00	Intensive diamond digging observed in RUF-controlled areas of Sierra Leone, coinciding with dry season

Despite the slow pace of the peace process, even its near collapse in May, during 2000 there was an apparent steady improvement in business in Bo and Kenema. The main highway to Bo and Kenema was open to commercial traffic for most of the year, and the visual signs of painted storefronts and re-stocked shops are reflected in the following diamond export statistics.

Table 3. Sierra Leone Diamond Exports

Dates	Carat Weight	Value in Us$	Average $ per Month
Jan-Jun 1999	2,319.03	$ 325,029.76	$ 54,171.63
Jul-Dec 1999	7,000.97	$ 919,795.58	$ 153,299.26
Jan-Dec 1999	9,320.00	$ 1,244,825.30	$ 103,735.44
Jan-Jun 2000	26,331.63	$ 3,448,336.94	$ 574,722.81
Jul-Dec 2000*	50,281.51	$ 6,533,134.70	$ 1,088,855.70
Jan-Dec 2000	76,613.14	$ 9,981,471.60	$ 831,789.30
Jan-Mar 2001	48,926.40	$6,586,385.00	$ 2,195,461.60

*includes the export of the "stockpile" of diamonds accumulated by some exporters during the Jul-Oct 2000 period under UN sanctions, and exported in Oct 2000 once the GOSL received an exemption
Source: GOSL, Government Gold and Diamond Office

Export statistics from the last two years show two trends. First, the increase in official exports from 1999 to 2000 clearly shows the return of some license holders. (There are not yet any exports from any of the new groups licensed in December 2000.) There were no official exports in 1998. Exports in 2000 were eight times greater than exports of 1999.

Second, the statistics from November 2000 through March 2001 show significant increases in diamonds through the official channel. A simple extrapolation from the January 2001 export level would put the annual figures for 2001 at $14.6 million. An extrapolation from the January-March 2001 export level would put the annual figures for 2001 at $26,3 million. It should be noted that these official exports are the "clean" channel, with the safeguards described above to prevent conflict stones from getting in. There are seasonal adjustments that could be made, however recent data is so distorted by other factors that the basis for a seasonal correction cannot be discerned. Anecdotal evidence indicates that during the dry season there is more digging and less sifting (to find rough stones) than in the rainy season.

The recent export performance in the GOSL clean channel raises a large question mark concerning current levels of smuggling of diamonds from RUF-held areas. The RUF-held areas on average possess the highest valued rough diamonds, especially Tongo Field and the Kono area. Using the extrapolation from January-March 2001, if the "normal" annual market value of Sierra Leone's exports were about $70 million (as estimated by international industry sources), that would imply that the Kono and Tongo Field diamonds are worth about $43.7 million per year.[2] Either these stones are not being marketed at the present time, or else there is a lot of smuggling occurring.

Thus, we are back to the opening statements of this report: it's about smuggling. The clean channel of exports from Sierra Leone is in operation, and it may be clean of conflict stones. There are anecdotal reports of diamonds from RUF-controlled areas being brought to Kenema and Bo. If they are being sold, those transactions are in contravention to Sierra Leonean laws and government regulations. If they are being exported, they are being smuggled, and the importers of these diamonds are in contravention to UN sanctions, and are circumventing the World Diamond Council's decision to permanently expel any such dealers from all bourses, worldwide.

MONITORING BY THE GOSL

Monitoring of "conflict" diamonds inside of Sierra Leone involves two different approaches:

1. Monitoring the diamonds in the "clean" channel to identify "conflict" stones (those that originate from areas under rebel control);
2. Monitoring the transactions in the chain of custody, from the mining field to export, to identify smuggling.

The first of these tasks is being done with a fairly high degree of confidence. As described above, the GGDO expert is backed up by the independent valuator, and they could be further backed up by surprise "audits" at the importing side. The surprise audits, both random and selected, could be done both by scrutiny by a panel of experts of the digital photographs of parcels sent from Freetown, and by physical examination of the goods, if there were suspicion of "conflict" stones in the parcel.

There is very little possibility that additional conflict stones, or any other stones, can be inserted into the parcels being shipped under the new

---

[2] These estimates are very soft numbers. The only hard numbers are the actual official exports.

Certification regime, provided that the importing authority confirms that all documentation is complete and correct, including the digital photographs; and that the security seals are in tact. At the present time, the likelihood that exporters are including conflict stones in the goods being exported with valid and correctly executed Certificates of Origin is small, given these mechanisms for detection of conflict diamonds in the "clean channel". Exporters who wish to continue to do business are cognizant of the scrutiny being placed on exports in the clean channel of official, GOSL-certified parcels.

Therefore, it is safe to say that conflict stones are not entering the clean channel, in any significant quantities, of rough diamonds exported under the official Certificate of Origin regime. The more likely scenario for the shipment of conflict stones out of Sierra Leone is that they are being smuggled.

Apart from RUF smuggling, the incentive to smuggle is now probably substantially related to tax evasion in the importing country. The export tax in Sierra Leone is reasonable, and it is probably being fairly administered. USAID/OTI began providing some technical assistance, in March 2001, by a fiscal expert to help estimate expected receipts from all sources related to diamonds, including the export fees and income from the schedule of fees shown in the attachment to this chapter.

The costs to, and risks of, smuggling in Sierra Leone have increased, given the scrutiny of the international community and the existence of UN sanctions. To detect smuggling now, monitoring mechanisms within Sierra Leone must be further developed.

Monitoring is an important task of public administration. If done correctly (following regulations and procedures), on a timely basis, in the right locations, and without corruption, it can both prevent illicit activity, and identify illicit behavior, in order to enforce the law.

The GOSL has licensing regulations on the books, but enforcement has been historically inadequate. Detection of illicit activity is in large part a responsibility of the mines monitors, whose job it is to monitor mining to detect smuggling, from the pits to the ports, and other major transit points.

OTI and MSI conducted two workshops on the subject of monitoring, one in December 2000 and one in February 2001. Their purpose was to obtain the active participation of public sector offices and individuals concerned with monitoring, and to inform key civil society stakeholders, and engage them in developing plans for better monitoring. A joint public-private effort is necessary to identify "conflict" diamonds at the source, as well as to reduce smuggling over the long term.

The GOSL has requested capacity-building for mines monitoring, including training of personnel. In addition, the GOSL would benefit from assistance to thoroughly review licensing and monitoring regulations and procedures, and assistance for the preparation of a manual for mines monitors. These are areas of on-going assistance from USAID/OTI, with MSI consultants.

Improvement of mines monitoring and revenue collection and allocation is being treated by the GOSL as an issue of public sector reform, with the active participation of the GOSL's anti-corruption unit, the Attorney General's office, and the Ombudsman. This is the public sector side.

Stakeholders

On the civil society side, probably the best information on who is smuggling diamonds is held by the producing communities themselves. It is the diggers and miners who know what stones are mined, by whom and where. It is the merchants in the diamond producing areas that hear about transactions, and see the effects of income being paid for diamonds.

If the communities themselves had an incentive for making sure diamonds mined in their area were declared, and marketed through the official, clean channel, then they might be forthcoming with information on mining, and information on smuggling. One or two people can be silenced with bribes, but entire communities cannot be bought off by smugglers.

Disclosure of diamond transactions is one of the necessary conditions for a (legally) functioning market system. Disclosure is necessary for tax collection and to identify smuggling. It is the diggers, their households, their extended families, their neighbors, and the traditional authorities that know what really happens above and below the surface of society. It is these people, the ultimate victims of war or beneficiaries of peace, that hold the key to disclosure. The mechanisms to reinforce silence and passive complicity have been stronger than the rewards for disclosure. The key to disclosure is to build in incentives for it, by those who have the information.

Incentives and Disincentives

An effective instrument against tax evasion and corruption is for there to be a balance of stakeholders inside of a system that will identify corruption, by having a stake in correct practices. The collection of taxes needs to be linked as directly as possible to the receipt of benefits from those taxes by those people who have knowledge about who and how much taxes should be paid.

In December, 2000, the GOSL, by a decision of Cabinet, approved the "earmarking" of .75% of the value of exports, to be allocated to diamond producing communities. The new distribution of the total 3% export tax is as follows:

0.75 %	to the GGDO to cover costs of GOSL valuation and export processing
0.25 %	to the GOSL/Ministry of Mineral Resources for monitoring
0.50%	to the independent international valuator
0.75 %	to a new Community Development Fund for diamond-producing communities
0.75 %	to the general GOSL Treasury
3.00%	*Total*

Prior to this new distribution, the GOSL export tax was also 3%. The low level of exports in 1999 and 2000 implies tax collection was a small amount. Total collections from the 3% export tax in 1999 were US$ 37,346.00 A comparison of the "old" (prior to the December 2000 Cabinet decision) and new distribution of the export tax is as follows:

Table 4. Comparison of the Distribution of Diamond Export Taxes Under Old and New Policies

Recipient of export tax	Old Policy	New Policy
GGDO: valuation	1.00 %	.75 %
Independent international valuator		.50 %
Ministry of Mines for monitoring	.50 %	.25 %
General Treasury	1.50 %	.75 %
Community Development Fund		.75 %
Total	3.00%	3.00 %

ORDER OF MAGNITUDE OF THE TAX BENEFITS OF LEGITIMATE DIAMOND OPERATIONS

Taxes

The following table presents projected annual receipts from diamond export taxes, for different levels of total exports.

Table 5. Projected Annual Receipts from Diamond Export Taxes

Tax recipient	Exports of: $5 million	Exports of: $10 million	Exports of: $30 million	Exports of: $50 million	Exports of: $80 million
GGDO	$37,500	$75,000	$225,000	$375,000	$600,000
GOSL for monitoring	$12,500	$25,000	$75,000	$125,000	$200,000
Independent Valuator	$25,000	$50,000	$150,000	$250,000	$400,000
Community Development Fund	$37,500	$75,000	$225,000	$375,000	$600,000
GOSL Treasury	$37,500	$75,000	$225,000	$375,000	$600,000
Total	**$150,000**	**$300,000**	**$900,000**	**$1,500,000**	**$2,400,000**

The allocation to the producing communities achieves two objectives. One objective is the devolution of value to the community at large, to be used for development purposes. Some might argue that this is compensation for environmental damage done by mining. There is a fee paid by the miner, additional to the land-use fee, which is supposed to pay for land reclamation. Diamond mining destroys the land for farming, as the thin layer of top soil is removed. Even if the pits were refilled, the top soil is lost in the careless, haphazard process of digging. The only way to reclaim the land is to carefully separate the top soil before digging begins, so that it can be replaced (on top), once the digging is finished. Given the primitive techniques being used at the present time, it is safe to say that no mining is being done with adherence to methods that would allow environmental reclamation. The advantage to large-scale, commercial mining is that it is relatively easier to impose standards of land use. However, it is not impossible, with a program of environmental education, for small-scale miners to adhere to practices of digging, especially the conservation of top soil, that will permit land reclamation. This will be part of the next stage of OTI's diamond development program up country.

The fees paid for environmental reparation are insufficient for the GOSL to repair environmentally wasted land. Top soil cannot be bought for this price. The fees paid are probably enough to pay for environmental education campaigns, and for monitoring of correct digging practices, that will permit land to be reclaimed for agriculture.

The objective of the new earmarking of export taxes for the Community Development Fund is to make the mining communities stakeholders in the correct operation of the "clean" channel of exports. There is a direct link between the value of Sierra Leone's official diamond exports and the money that diamond-producing communities will receive. On February 6, 2001, the GOSL began depositing the .75% in a special account. OTI and MSI facilitated a workshop on February 7-8, 2001, to inform a small group of representatives from the diamond-producing chiefdoms, and from Bo and Kenema, the two urban centers in the diamond-producing areas, about the new tax ear-mark, and in general about the issue of monitoring.

An important task, from the point of view of effective monitoring, is how to engage the communities to report diamond production in such a way that the information can be compared with export data, and also serve as the basis for the allocation of earmarked export revenues among the various producing chiefdoms.

The Bank of Sierra Leone has indicated they will publish data on official exports on a regular basis, such as once per month. The producing communities will be able to track the value of exports, and track the value of the expected tax revenue that should be paid into their Diamond Development Account, for direct distribution to their communities.

During the February workshop, participants proposed four possible models for an allocation mechanism, and these will be further discussed in March. The GOSL is expected to adopt an allocation formula developed with the participation of a broad representation of public and private sector "stakeholders".

MARCH 2001 WORK PLAN

From March 19 to April 6, 2001, USAID/OTI with Management Systems International consultants implemented the following work plan in Freetown.

1. Training for Mines Monitoring Officers

The training of the mines monitoring officers focused on search procedures, ethics, record keeping, communications protocols, and performance criteria. In addition a full review of mining regulations and the duties of monitors was to be done. A manual of monitoring operations will be developed for their field reference.

2. Revenue Receipts and Allocation

Technical assistance was provided to develop procedures for estimating, tracking, reporting, and auditing the fiscal receipts from mining licenses, monitoring fees, and export taxes. The accounts reporting needs to be transparent and easily understood by representatives of the civil society, and the diamond-producing communities.

Technical assistance was continued to help the GOSL design a system for allocating the .75% export tax ear-marked to producing communities. This was started in the workshop in February, 2001, as described above, with four models proposed to accomplish the objectives of equity in resource allocation at the same time that the tax distribution serves as an incentive to communities to help identify and prosecute smuggling and "conflict" diamonds.

3. Public Information

A public information process needs to be strengthened, including workshops to inform the public in Bo and Kenema on the new diamond policies and the community development fund, and to inform the public about the efforts to monitor mining to prevent smuggling and control "conflict" diamonds. These workshops could be carried out jointly by the GOSL with civil society organizations. It is also appropriate for civil society organizations to conduct their own information and sensitization activities.

OTI can support the dissemination of information through support to workshops as well as support for the production and airing of radio information. A preliminary proposal for a Public Information Campaign was received by OTI. It is being reviewed, and some of the activities proposed may be funded.

4. Equipping the GOSL Mines Monitors

Once the mining sector is fully reformed, and the fiscal receipts are being collected and correctly applied, the GOSL will have enough resources for a sustainable regulatory function. However, in order to jump-start the monitoring operation, which is critical to the detection of "conflict" diamonds and smuggling, OTI may assist the GOSL to equip the mines monitors, such as with equipment for the production of laminated picture IDs for all mines monitors, and for all dealers, agents, and exporters.

NOTE ON OTI'S METHODOLOGY

This paper has reported on USAID/OTI's assistance to the Government of Sierra Leone since December 1999 to address the link between diamonds and the rebel war. A chronology of actions and assistance provided by OTI with its consultants from Management Systems International is annexed. A number of international initiatives have developed simultaneously with this effort in Sierra Leone, including United Nations sanctions on conflict diamonds from Sierra Leone, and the creation of the World Diamond Council to coordinate industry response to this issue. International initiatives are very important to success in curbing conflict diamonds in Sierra Leone. However, the Sierra Leone program is being implemented on the assumption that international measures might support success, but should not be relied upon to solve internal problems.

The GOSL with USAID assistance has approached conflict diamonds as a complex development problem of the Sierra Leonean society: its traditional economy and contemporary business sector, government institutions, and the socio-economic systems that perpetuate abject poverty in areas endowed with alluvial wealth ready for the taking.

USAID/OTI has acted as a consultant. A consultant does not take charge of implementation; he or she recommends to the principles, and provides a longitudinal effort of diagnosis, recommendation, re-evaluation, and new recommendations. This has been OTI's approach, working with MSI. This approach has some disadvantages, but the biggest advantage is that the resulting new policies are genuinely the product of Sierra Leonean action, therefore they are likely to have permanence.

In summary, it is the monitoring program that now requires the most attention, to complete Sierra Leone's program of control of "conflict" diamonds. Most of the new diamond policy has been adopted. The

Certification system is in use. In particular, the new foreign exchange requirements, and the installation of an independent valuator are important components to the "clean" channel.

As noted at the beginning of this report, enforcement mechanisms are beyond the scope of USAID/OTI's current work in Sierra Leone. The judicial framework is important once executive orders begin to be enforced. Mechanisms for receipt of complaints need to be developed. Confiscation needs to be done according to the rule of law, and remedies need to be judiciously and expeditiously applied. The engagement of the Ministry of Justice, the Office of the Ombudsman, and the GOSL's special anti-corruption office will be necessary as the development of better mechanisms for monitoring create a demand for law enforcement.

NOTE ON THE CMRRD AND THE RUF

In July 1999, the Government of Sierra Leone (GOSL) and the rebel Revolutionary United Front (RUF) acknowledged in Article VII of the Lome Peace Agreement that diamonds are crucial to war and peace in Sierra Leone. The Lome Agreement provided for the creation of a new Commission for the Management of Strategic Minerals, National Reconstruction and Development (CMRRD), and provided a post-conflict role for the RUF, to hold appointed office in this new government para-statal organization, that would develop new diamond policies. At the time of the peace agreement, the RUF controlled some of the most important diamond-producing areas in the country. They still do. Despite adherence to a cease-fire in Sierra Leone, the RUF have not disarmed and demobilized, and have not granted free and full access to GOSL authorities and UN peace-keepers into RUF-held territory.

Diamond development initiatives have proceeded without the CMRRD. Once the GOSL has full control over all diamond-producing areas, and if RUF combatants have voluntarily disarmed and demobilized, the GOSL will have the prerogative to grant mining licenses to former combatants, who pledge to reintegrate as demobilized ex-combatants. Hundreds of ex-combatants are pursuing new livelihoods today in many areas of Sierra Leone. Reconciliation and reintegration will have to be achieved in the diamond-producing areas currently under RUF control, as well. Through the new ear-mark of export taxes, the GOSL will be supporting communities in their development with the new diamond development funds.

ACRONYMS

GOSL	Government of Sierra Leone
NGO	non-governmental organization
GGDO	Government Gold and Diamond Office
BOSL	Bank of Sierra Leone (Central Bank)
HRD	Diamond High Council of Belgium
USAID	United States Agency for International Development
OTI	Office of Transition Initiatives
MSI	Management Systems International
TA	technical assistance
CO	Certificate of Origin
WDC	World Diamond Council
UN	United Nations
US/UN	United States Mission to the United Nations
US	United States
UK	United Kingdom
RUF	Revolutionary United Front (Sierra Leone rebels)
UK/FCO	United Kingdom/Foreign and Commonwealth Office
CMRRD	Commission for the Management of Strategic Resources, National Reconstruction and Development
USAID/OTI	US Agency for International Development/Office of Transition Initiatives

ANNEXES – AVAILABLE ELECTRONICALLY OR BY FAX, AS INDICATED

1. Electronic: USAID/OTI Questions and Answers on "Conflict" Diamonds, 03-15-01
2. Electronic: Chronology of Actions and Assistance of USAID/OTI to April 6, 2001
3. FAX: Map of Sierra Leone: UN Blue Line to March 15, 2001 and major diamond areas in production
4. FAX: Certificate of Origin Number 000001
5. FAX: Procedures for the Issuance of Diamond Export Licenses by the Government of Sierra Leone
6. FAX: Mines and Minerals Act, Artisanal/Small-Scale Mining Licence, Application for Precious Mineral Exporter's Licence
7. FAX: Schedule 1: Revised Fees for Small-Scale and Artisanal Mining and Marketing under the Mines and Minerals Act
8. FAX: Agreement between the Government of Sierra Leone and Diamond Exporters
9. FAX: Procedures for the Valuation of Diamonds by the Government Gold and Diamond Office (GGDO)
10. FAX: Banking Guidelines for Diamond Exporters
11. FAX: United Nations Security Council Resolution 1306 (2000)

USAID OFFICE OF TRANSITION INITIATIVES (OTI) Q'S AND A'S ON "CONFLICT" DIAMONDS: SIERRA LEONE

1. Is there, at the present time, a "clean" channel of diamond exports from Sierra Leone?

Yes, there is a channel for official, conflict-free diamonds from Sierra Leone, but only to Antwerp at the present time. UN Resolution 1306 (2000) of July 5, 2000, prohibited the import of Sierra Leone diamonds unless the Government of Sierra Leone (GOSL) had an operating Certification of Origin system. In October 2000, the UN granted the GOSL an exemption, to restart diamond exports using a new certification system that was developed with technical assistance from a trilateral team (US, Belgium, and UK), mostly provided by the Diamond High Council of Belgium and USAID's Office of Transition Initiatives. The UN-approved certification system consists of the Certificate of Origin document, plus an electronic tracking and confirmation system with digital photographs. There is a chain of authorized transactions from the exporter back to the diamond digger, however, the GOSL does not have the historical practice or the current means to adequately monitor all of the transactions. More work on monitoring has to be done before we can assert that conflict stones are not entering this clean channel.

The reason this channel is only open to Antwerp is because part of the new Certification of Origin system involves electronic tracking of exports and imports, digital photographs of parcels, and other procedures to ensure that the goods certified at origin are the same goods that are officially imported at destination. The system is not complicated. It should be adopted and installed by all importing countries, but is currently only in use in Antwerp.

2. Is it better to stop the exports of all diamonds from Sierra Leone, until an adequate system for monitoring can be put in place?

To prohibit exports (continuation of sanctions) and thus stop the flow of diamonds through the official (as yet imperfect) channel would be counter-productive, as it would encourage the continuation of smuggling. Smuggling is allegedly widely practiced or tolerated in the international diamond trade, and is the most important obstacle to a "clean" channel. Smuggled diamonds have accounted for a large proportion of Sierra Leone's output for several years, and almost 100% in 1999. "Conflict" stones are mixed into the flow of smuggled goods. Without controlling smuggling it is impossible to control 'conflict' diamonds. Therefore, resuming legitimate, declared exports is fundamental to getting control over "conflict" stones. It is a necessary, if not sufficient, condition to solving the problem.

3. Are there any indicators that smuggling has been reduced out of Sierra Leone since the new Certification of Origin regime went into effect?

Yes, there are several indicators. 1. Diamond industry experts consulted in Conakry and Freetown suggest that a reduction in recent exports from Guinea approximately matches the trend of increased exports from Sierra Leone. 2. There are anecdotal reports that known traders are not as willing to purchase parcels of stones of questionable origin, and some dealers are having problems selling some parcels in Guinea and Liberia; and there are some reports of lower prices in these markets. 3. Official exports from Sierra Leone have increased. 4. Diamond exporters holding licenses before sanctions went into effect declare that they were holding stones, waiting for a legitimized new regime. This constituted the so-called "stockpile", and there are many reports in Freetown about pressure to export stones before the new certification regime went into effect on October 27. Official exports were suspended from July 6 to October 27. Delays in authorizing some Certificates of Origin, in late November, after the first dozen certificates were issued, are reported to have encouraged some exporters to return to the use of illegal channels.

4. Are there any indicators that "conflict" stones are coming through the official channels?

As of the first week of February, 2001, the GGDO twice had flagged possible stones whose origin is RUF-controlled areas, since exports restarted on October 27. This was twice out of about 22 parcels presented for certification. The GGDO valuator is an expert, said to be able to detect the origin of individual stones, as well as any world-class valuator can. The only ways to improve this detection at the point of export would be to have an additional expert (independent valuator) look at the stones and to have better monitoring of all transactions from the field to the point of export. In August 2000, the GOSL requested an independent valuator be provided by the international community. In December 2000, the GOSL approved a modification of diamond export taxes to allocate monies to pay for an independent international valuator. A Belgian firm recommended by the Diamond High Council (HRD) was contracted and, the first week in February 2001, fielded a full-time expatriate resident expert in Freetown, to perform independent valuation of all parcels presented for export to the Government Gold and Diamond Office (GGDO). OTI is providing training and some equipment to help the GOSL improve monitoring, engaging civil society and diamond-producing communities as stakeholders in monitoring.

5. If and when the RUF no longer control the diamond mining areas, what mechanisms will be employed to ensure that revenues are used for legitimate purposes and not simply revert to corrupt officials?

USAID's Office of Transition Initiatives (OTI) is implementing a comprehensive program of assistance to the Government of Sierra Leone (GOSL) on diamond policy and development. As part of this process of reform, the GOSL has made two changes to the export tax on diamonds. First, they have ear-marked funds to pay for an expatriate independent diamond valuator, as oversight to their own export valuation and to detect "conflict" diamonds that could be introduced into the clean channel of official exports, which began to operate in October 2000 under the new certification of origin regime. Second, they have ear-marked funds to be allocated to diamond-producing areas, to encourage greater involvement of communities in monitoring for conflict diamonds, and to assist with development efforts in these war-ravaged areas. USAID is providing fiscal and governance experts who are working with the GOSL, and with community representatives and NGOs, to create transparent and accountable mechanisms to manage these monies, and to monitor all diamond transactions.

SIERRA LEONE: DIAMOND POLICY AND DEVELOPMENT PROGRAM CHRONOLOGY OF ACTIONS AND ASSISTANCE OF USAID/OTI TO APRIL 6, 2001

- *Technical assistance to Lome Peace Process* (April-July '99): work with civil society and the Government of Sierra Leone (GOSL) to identify and address causes of conflict, including illicit exploitation of diamonds by the rebel RUF
- Oct '99: US Secretary of State announces *$1 million of USAID/OTI assistance* to the Commission for the Management of Strategic Resources, National Reconstruction and Development (CMRRD) and other peace structures
- *Jan 2000: Technical Mission to Freetown* to discuss CMRRD with GOSL and establish conditions for implementation. Basic pre-condition for beginning CMRRD: substantial progress on disarmament, demobilization and reintegration (DDR) and free access of GOSL and civilians to diamond-producing areas
- *CMRRD Strategic Planning Exercise in Freetown, March 20-24, 2000*, with diamond industry leaders and experts; De Beers proposal follows (letter of May 4, 2000, to State Department with proposals for how to identify conflict diamonds)
- *Kimberley, South Africa Technical Forum* on "Conflict" Diamonds, May 11-12, 2000: OTI participation with State Department

Sanctions on UNITA Diamonds (Excerpt from UN Security ... 139

- *OTI Working Paper: 05-08-00 "Proposal for a New Diamond Policy and Operations"*: GOSL publishes under title "Guidelines on the Mining and Marketing of Diamonds in Sierra Leone"
- *Lwanda Technical Meeting*: (USG not present but working group incorporates part of OTI Working Paper into Lwanda report)
- *London Consultations*, June 15-16, 2000: UK/FCO, Diamond High Council, Belgian Ministry of Economic Affairs, De Beers, USG
- *Freetown Trilateral Mission*, July 14-15, 2000: US/UK/Belgian mission
- *World Diamond Congress*, July 17-18, 2000, Antwerp (OTI attends as observer)
- *Technical Meetings with GOSL and HRD to finalize Certification of Origin regime* for Sierra Leone: Antwerp (July 18), Washington (July 27) and New York (July 28, "preview" meeting at UN with Sierra Leone, UK, US, Belgium, Israel, India)
- *Technical Assistance to GOSL, with HRD, for Sanctions Committee presentation*, July 2000
- *UN Sanctions Committee Hearings*, July 31 – August 1, 2000, technical support with US/UN Mission, New York
- *Tel Aviv- World Diamond Council, inaugural meeting* September 7-8, 2000, OTI attended as observer for USG
- *Technical Assistance Mission to Freetown, September 11-15*, continue work on Certification of Origin regime and policy development, as per September Assistance Plan
- *Technical Assistance Mission to Freetown, October 23 - November 1, 2000* by Management Systems International (MSI): consultants to USAID-OTI
- *Technical Assistance Mission to Freetown, December 6-15, 2000*, review of start-up of certification regime; discussion of export tax revenues, banking and foreign exchange policy and other pending policy as per "Guidelines" policy document; workshops on monitoring of "conflict" diamonds, including GOSL, traditional authorities, NGOs and other civil society groups
- *Technical Assistance Missions to Freetown, January 22-26 and February 5-9, 2001*, workshops and policy development on monitoring,

especially mechanisms for allocation of Community Development Fund (ear-mark of export taxes) and mechanisms for community participation, transparency and accountability in Fund management

- *Technical Assistance Mission to Freetown, March 19-April 6, 2001*, further development of monitoring, including training of GOSL Mines Monitoring Officers, development of information campaign for miners and the general public, development of support for community engagement in monitoring for "conflict" diamonds and in ear-marked export tax revenues

- 2000-2001: Participation in various academic, NGO and US Government meetings/conferences to develop understanding and advocacy on "conflict" diamonds and Sierra Leone: INR/Meridian International Center; International Peace Academy; White House Conference on Diamond Technologies; InterAction

INDEX

A

acceptable standards, 25, 107
accountability, 82, 84, 92, 93, 97, 100, 108, 140
accredited export agencies, 16
Afghanistan, 7, 8
Africa, 73
African conflicts, 99
al Qaeda financial dealings, 10
al Qaeda operatives, 8, 9
al Qaeda, 6-10
Albright, Secretary, 99
Alliance of Democratic Forces for the Liberian of Congo-Zaire (ADFL), 37
Alluvial deposits, 48
Alternative Investment Market (AIM), 39
Amnesty International, 65, 106
Angola, 2, 4, 6, 7, 17-19, 21, 24, 27-29, 31-37, 42, 43, 53-57, 59, 61, 62, 66, 73-75, 76, 81, 83-85, 87, 91, 99, 101, 105, 114
Angolan conflict diamonds, 5, 36
Angolan diamonds, 32, 35, 36
Angolan government, 15, 33, 35
Annan, U.N. Secretary General Kofi, 34
annual diamond production, 4, 48

Antwerp Resolution, 17
Antwerp, 15-17, 36, 43, 49, 66, 75, 87, 104-106, 115, 116, 119, 120, 135, 136, 139
armed conflict, 2, 24, 54, 73, 87, 103, 106
armed conflicts, vii, 2
armed forces, 43, 66
armed hostilities, 29, 32
arms and diamond embargoes, 78
arms brokering, 50
arms deals, 82, 83
arms manufacturers, 75
arms trading, 34
arms trafficking, 13, 22
Artisanal mining, 48
artisanal, 4, 33, 36, 37, 39, 43
Ascorp, 17, 35, 36
audit(ing), 74, 95, 97, 118, 131
audits, 96, 125

B

Bah, Ibrahim, 8
Belgium, 7, 9, 15, 16, 24, 25, 32, 34, 37, 49, 50, 57, 66, 68, 78, 89, 100, 105, 112, 113, 116, 134, 135, 139
bin Laden, Osama, 6, 7, 10
bin Laden, Osama, 6, 9
Black Caucus, 70, 71

black market, 44, 48, 52, 53
blood diamonds, 2
bombing, 6
Botswana, 10-13, 19, 24, 29, 30, 57, 68, 70, 86, 100, 101
burdensome regulations, 89
Burkina Faso, 8, 34, 47, 48, 50, 66, 70
Bush Administration, 1, 22, 23, 69, 70
Bush, George W., 71
buy and sell, 52
buyers, 8, 9, 18, 36, 110, 112

C

Cameroon, 78
Carter, President James, 70
Cartier, 67
Central Bank, 115, 120, 121, 134
Central Selling Organization, 3
Centre National d'Expertise (CNE), 43
Certificate of Origin (COO), 15, 16, 20, 28, 32, 35, 49, 51, 52, 74, 76, 77, 84, 87, 91, 105, 113-118, 120, 126, 134-136
certification fee, 51
certification scheme agreements, 27
certification scheme, 26-28, 49, 50, 81, 87, 92, 93, 100, 101, 117
certification system, 17, 23, 25, 26, 31, 35, 43, 51, 69, 93, 100, 105, 135
certification systems, 16, 21, 24
child soldiers, 59, 61-63
Chile, 77
Chowdhury, Ambassador Anwarul Karim, 78
CIA, 66, 67
civil conflicts, 2
Civil Defense Forces (CDFs), 109
civil society, 75, 105, 112, 126, 127, 131, 137, 138, 139

Clean Diamonds Act, 30
clear-cut policy, 54
Clinton Administration, 1, 11, 19-22, 68
Clinton years, 67, 68
Clinton, President William J., 49
cold war, 66
commerce, 14, 19, 30, 53, 55, 83, 110
commercial activities, 10
commercial banking, 122
commercial fraud, 13
commercial invoice, 52, 91
Commission for the Management of Strategic Minerals, National Reconstruction and Development (CMRRD), 133, 134, 138
commodity prices, 44
Compaoré, President Blaise, 34
conflict diamond imports, 70
conflict diamond issue, 67
conflict diamond trade, 1, 2, 4-6, 12, 13, 21, 23, 24, 30, 50, 66, 67, 86, 97
conflict diamond-related legislation, 12
Conflict Diamonds Act, 31, 66, 69
conflict diamonds, 73, 74, 75
conflict resolution, 5
conflict stones, 18, 50, 104, 107, 119, 120, 124-126, 136
conflict-diamond countries, 68
Congolese diamonds, 37, 40, 42
consignment, 51, 90
consumer boycott, 11, 65
consumer rejection, 11
control environment, 84, 93
corruption, 41, 54, 67, 88, 107, 108, 110, 113, 119, 126, 127, 133
Côte d'Ivoire, 34
country of origin, 12, 84, 85, 92
crime, 82, 84, 89, 104, 110
criminal activity, 104
criminal elements, 88
criminal networks, 50

Index

currency revaluation, 120
currency transactions, 53
customs and immigration, 9
cutting and polishing, 68, 86, 90
cutting, 15, 65, 67, 68, 86, 90

D

Dar es Salaam, Tanzania, 9
De Beers Diamond Trading Company (DTC), 3, 18
De Beers, 3, 4, 12, 18, 35, 36, 55, 65, 67, 86, 100, 101, 105, 138, 139
dealers, 7, 9, 16, 34, 109, 110, 112-114, 121, 125, 132, 136
Debswana, 12, 13
deep mining, 86, 88
Democratic Party, 67, 70
Democratic Republic of Congo (DRC), 2, 4, 16, 17, 34, 31, 37-45, 53-55, 59, 63, 64, 66, 70, 81, 83, 85, 89
detection of "conflict" diamonds, 132
detection of smuggling, 123
diamond areas, 37, 43, 135
diamond buyers, 8, 11, 36, 122
diamond commerce, 15, 16
diamond cutter, 67
diamond export sales, 29
diamond export taxes, 129, 137
diamond exporting countries, 74
diamond exports, 85, 86, 89, 100, 104, 106, 114, 121, 130, 135
diamond flows, 83, 89
Diamond High Council (HRD), 15, 16, 24, 32, 33, 37, 49, 51, 56, 100, 105, 113-115, 119, 120, 134, 135, 137, 139
diamond imports, 3, 4, 16, 30, 56, 82, 83, 84, 90, 91
Diamond Industry, 26, 35, 40, 86
diamond industry, vii, 1, 10, 11, 14, 15, 19, 20, 22, 23, 25, 26, 35, 36, 40, 42, 43, 53, 55, 65, 74, 75, 78, 81, 83, 86, 90, 99-101, 103-106, 138
diamond jewelry, 3, 11, 69
Diamond Manufactures Association, 52
diamond market, 3, 18, 45, 110
diamond mining, 3, 4, 7, 8, 23, 35-37, 39, 40, 43, 46, 85, 101, 137
diamond pipeline, 89, 94
diamond producing countries, 1, 11, 20, 86
diamond producing state, 19
diamond production, 48
diamond sales, 35, 52, 65, 66
diamond sanctions, 75
diamond trade, 1, 11, 15, 16, 19, 21, 22, 29, 34, 42, 53, 54, 88, 90, 92, 101, 103, 104, 121, 136
diamond traders, 7, 17, 36, 43, 44, 75, 122
Diamond Trading Company Limited, 18
diamond trading, 8, 11, 14, 16, 20, 28, 34, 40, 43, 50, 52, 54, 55, 86
diamond traffic, 103
diamond transactions, 9, 15, 54, 82, 84, 113, 120, 127, 138
diamond wealth, 2
diamond-cutting factories, 67
diamond-producing areas, 106, 109, 111, 115, 130, 133, 138
diamond-related investigations, 91
diamonds, non-industrial, 4
diamond-trading houses, 44
diggers, 86, 88, 108, 109, 112, 121, 122, 127
digital photographs, 114-116, 125, 126, 136
disarmament, 7, 77, 138
disarmament, demobilization and reintegration (DDR), 138
domestic diamond industries, 20
domestic diamond marketing, 23

Dos Dantos, President Jose Eduardo, 32
draft bill, 69
drug deals, 89

E

Economic Community of West African States (ECOWAS), 5, 22, 77
economic development, 75
economic transparency, 44
electronic tracking system, 115, 116
embargo, 22, 32, 76, 78
EU members, 26, 48
Europe, 68, 85, 101, 122
European Union (EU), 5, 26, 27, 48, 56, 86, 96
exchange rate, 45, 120
exploitation, 19, 41, 105, 110, 138
export and import control system, 18
export and import data, 89
export control(s), 15, 16, 23, 30, 51, 54, 55
export diamonds, 44, 111
export licenses, 112, 135
export licensing, 112
export of arms, 22, 50
export of diamonds, 31, 48, 50, 105
export tax(es), 126, 128, 129, 131, 137, 139, 140
exporters, 91, 108, 110, 113, 114, 117, 120-122, 124, 126, 132, 136
Export-Import Bank (Ex-Im Bank), 30, 31, 67
exporting authority, 51, 114, 115
exporting, vii, 15, 17, 23, 30, 34, 51, 52, 74, 83, 84, 86, 91, 93, 104, 111, 113-115, 118
external monitoring, 27, 28, 29

F

faceted, 4

Federal Bureau of Investigation (FBI), 6, 7, 9
Federal Trade Commission, 88
fees, 42, 43, 53, 112, 113, 115, 119, 121, 126, 130, 131
financial crime(s), 104, 123
financial institutions, 75
financial intermediation, 103, 121
foreign buyer, 27
foreign currency, 43, 44, 52
foreign exchange market, 110
foreign exchange, 48, 107, 108, 115, 120-122, 133, 139
foreign investment, 45
foreign military support, 38
Fowler Report, 5, 6, 21, 33, 34
free market principles, 20
Freetown, 67, 100, 104, 105, 110, 113, 115, 116, 119, 122, 123, 125, 130, 136-140
fundamentalist Islamic group, 10

G

Gaborone, Botswana, 2, 28
Gambia, 66, 85
gem quality stones, 32
gem-quality diamond sales, 18
gemstone diamond production, 100
gemstones, 99
global certification scheme, 100
global diamond trade, vii, 53
global export certification system, 16
Global Witness, 65, 106
goods and services, 17, 45
government corruption, 45
Government Gold and Diamond Office (GGDO), 48, 110, 114-121, 124, 125, 128, 129, 134, 135, 137
government of national unity and reconciliation (GURN), 32
Government of Sierra Leone (GOSL), 48-52, 55, 77, 104-106, 108, 111-120, 122-135, 137-140

guerrillas, 66, 71
guidelines, 69, 93, 97
Guinea, 16, 48, 60, 66, 85, 87, 123, 136

H

hard currency, 48, 52
Harmonized Tariff Schedule (HTS), 4
health care, 13, 60
HIV/AIDS, 11, 13
Hizballah militia, 7, 8
hostilities, vii, 2, 45, 59, 77
human rights, 5, 11, 30, 60, 65, 68, 106
humanitarian crises, 81, 83
humanitarian spending, 59
Humanitarian Spending, 61, 62, 64

I

IDI Diamonds, 42
illegitimate diamond trade, 99
illicit activity, 104, 126
illicit arms, 22, 99
illicit behavior, 126
illicit dealings, 8
illicit diamond profits, 99
illicit diamond trade, 6, 13, 50
illicit diamonds, 12, 19, 22, 30, 74, 75
illicit sellers, 14
illicit trade, vii, 20, 22, 31, 50, 81, 83, 88
illicit trading, 6, 9, 19, 53, 54
import ban, 69
import confirmation certificate, 51
import control system, 84, 90
import controls, 18, 90, 91
import tax, 48
import/export, 17, 23, 25, 26
importing countries, 17, 20, 68, 115, 136

imports of diamonds, 31
incentives, 53, 55, 106, 107, 127
individual diamonds, 14, 15, 23
industrial grade diamonds, 32
intermediary buyers, 36
internal controls, 28, 92-95
internally displaced persons (IDPs), 59, 60, 62, 63
internally traded diamonds, 16
international advocacy campaigns, 5
international certification scheme, 24-26, 87
international certification system, 17, 27
international certification, 17, 24-27, 87, 93
international criminal activities, 13
international diamond certification scheme, 82-84, 92
international diamond cutting centers, 15
international diamond firms, 90, 120
international diamond industry, 13, 83, 86
International Diamond Manufacturer's Association (IDMA), 16, 17, 106
International Diamond Manufacturers, 87, 100
international diamond trade, 21
international financial crimes, 123
international governmental organizations (IGOs), 5
international law, 24, 54
international policy making, vii
international press, 10, 40, 47
international sanctions, 1, 19, 84
International Trade Commission, 4
international trade, 2, 23, 26, 27, 29, 35, 52-54, 69, 92
Islamic militancy, 10
Israel, 8, 17, 24, 43, 66, 68, 112, 115, 116, 139
Ivory Coast, 16, 60, 66

J

Japan, 101
Jewelers of America (JA), 12, 17, 67, 68, 70
jewelers, 12, 69, 71
jewel-quality gems, 37
jewelry, 3, 66, 69, 95, 104
jewelry-store owners, 71

K

Kennedy, President John F., 70
Kimberley Certification Scheme, 92
Kimberley member country, 26
Kimberley Process certificates, 85, 92
Kimberley Process, 2, 4, 5, 12, 17, 21, 23-30, 81-84, 86-88, 92-94, 96, 97
Kimberley-Plus Process, 24
kimberlites, 48

L

labor costs, 86
laser-scanning technologies, 15
law enforcement, 15, 36, 107, 122, 133
Lazare Kaplan International, 43, 65, 67
Lebanese Amal militia, 7
Lebanese Diaspora, 7
legal seal, 51
legitimate diamond industry, 87
legitimate diamonds, 10, 12, 30, 73, 74
legitimate international market, 14
legitimate trade, 24, 49, 50, 53, 54, 94
legitimately produced diamonds, 11
Liberia, 7, 8, 9, 16, 22, 31, 45-48, 50, 60, 66, 70, 75, 78, 82, 84, 85, 91, 123, 136

Liberian government, 9, 19, 22, 48, 78
Libya, 8
licenses, 42, 44, 111, 112, 131, 133, 136
licensing regulations, 126
licensing, 107, 108, 111
local diamond dealers, 44, 54
Lome Peace Agreement, 77, 123, 133
Lusaka peace process, 32
Lusaka Protocol(s), 32, 76

M

Management Systems International (MSI), 104, 126, 127, 130, 132, 134, 139
mandatory worldwide sanctions, 22
market demand, 3, 30
marketing of diamonds, 33, 53
marketing organizations, 10
marketing practices, 99
Member States, 74, 78
microscopic markings, vii
military actions, 5
military spending, 59, 61, 62, 64
mined diamonds, 8, 37
mineral wealth, 46, 59
miners, 36, 94, 112, 113, 127, 140
mines, 35, 40, 41, 43, 45, 50, 86, 90, 115, 126-128, 131, 132, 135, 140
minimal standards, 25, 28
minimum requirements, 26, 111
mining code, 43, 45
mining revenues, 86
mining rights, 3, 39, 45
mining, vii, 2, 15, 32, 35, 36, 38-40, 43, 45, 48, 53, 55, 67, 68, 75, 86, 90, 104, 105, 108, 109, 111, 112, 113, 115, 123, 125-127, 129-133
Ministry of Mineral Resources, 112, 121, 128
money laundering, 13, 82, 83, 89, 123

monitoring mechanism, 96
monitoring operations, 131
multi-lateral organizations, vii

N

Nairobi, Kenya, 9
Namibia, 10, 16, 19, 24, 25, 34, 37, 40, 68, 86, 100
national controls, 27
national diamond legislation, 24
national laws, 26, 54
National Patriotic Front of Liberia (NPFL), 46
national system of transparency, 24
National Union for the Total Independence of Angola, 82, 85
national unity, 77
natural resource exploitation, 5
natural resources, 3, 19, 41, 59
non-certified diamonds, 26
non-conflict producing states, 19
non-exclusionary system, 28
non-governmental organization(s) (NGO(s)), vii, 5, 11, 19, 22, 23, 25, 27, 29, 40, 41, 50, 55, 69, 75, 78, 83, 85, 87, 88, 97, 106, 108, 134, 138-140
Nujoma, Namibian President, 40

O

Office of Transition Initiatives (OTI), 103, 104, 126, 130-132, 134, 135, 137-139
official diamond market, 36, 37
official exports, 110, 117, 124, 125, 130, 138
official rate of exchange, 44
official sales and production, 4
open market competition, 23, 55
origin certification, 22
origin of diamonds, 12, 14, 15, 25, 43, 50

Oryx Diamonds, 39
Oryx Natural Resources, 39
Oryx Zimcon, 39
Overseas Private Investment Corporation (OPIC), 30, 31

P

Partnership Africa Canada (PAC), 14, 27, 56, 106, 110
peace agreement, 32, 133
peace, 73, 74, 75, 77
peacekeepers, 7, 46
peacekeeping, 46, 77
penalties, 16, 23, 26, 94
per capita income, 13
Petra Diamonds, 39
Physicians for Human Rights, 65, 69, 106
policy coordination initiatives, 1
policy issues, 14, 22
policymaking, 12, 21, 26
polished diamonds, 17, 67, 95
political tensions, 2
Popular Movement for the Liberation of Angola (MPLA), 32
poverty, 54, 60, 66, 103, 110, 121, 132
producing country, 25, 27
production capacities, 88, 90
production control system, 43
production statistics, 29, 84
profits, 8, 66, 67, 99
prosperity diamonds, 13
provenance of diamonds, 14, 42
public pressure, vii
publicity, 10, 11, 12, 40, 65

R

rebel groups, 65, 66, 73
rebel movements, 37, 81, 83
rebel officials, 41
rebel-held territory, 61, 62, 63

rebels, 7, 19, 59, 65, 75, 87, 104, 105, 123, 134
re-export, 25, 48
refugees, 59-61, 63, 66, 123
regional organizations, 78
remote areas, 82, 83, 88
required provisions, 27
retail industries, 10
retail jewelry market, 3
Revolutionary United Front (RUF), 7-9, 19, 22, 45-47, 49, 50, 65, 66, 73, 77, 78, 82, 84, 85, 91, 99, 104-106, 109-111, 123-126, 133, 134, 137, 138
risk assessment, 82, 84, 93, 94
river beds, 48
rough controls, 17, 68
rough diamond certification scheme, 2, 27
rough diamond export and import controls, 31
rough diamond import(s), 30, 31, 84
rough diamond parcels, 17
rough diamond production, 26, 89
rough diamond shipments, 26, 84
rough diamonds, 3, 4, 15, 17, 22-24, 26, 28, 35, 49, 73, 77, 79, 82-87, 89-93, 95, 100, 113, 114, 125, 126
rough diamond-trading firm, 42
RUF diamonds, 7, 8, 48
Russia, 24, 25, 32, 48, 67, 68, 86

S

sales, vii, 2
Security Council Sanctions Committee, 100
Security Council, 74-78, 84, 88, 99, 100, 105, 135
security forces, 9, 111
self-regulation, 26, 27, 95
Senegal, 77, 78
separatist Movement of Democratic Forces of Casamance (MFDC), 8

Sese Seko, President Mobuto, 34, 37, 67
shipments, 34, 84-86, 90, 91
Sierra Leone Political and Peace Council, 7
Sierra Leonean diamonds, 31, 49, 50, 53, 119
Sierra Leonean Gold and Diamond Office, 51
small businesses, 68
small-scale miners, 36, 113, 130
smuggling (of) diamonds, 110, 124, 127
smuggling, 9, 13, 36, 40, 42, 43, 49, 53, 54, 75, 88, 90, 91, 104, 106, 107, 110, 111, 113, 122, 124-126, 127, 131, 132, 136
social and educational institutions, 75
social disintegration, 11
Sociedade de Desenvolvimento Mineiro (SDM), 36
Societe de Development du Diamante (SDD), 42
socio-economic development, 10
sorting, 86, 104, 118
South Africa, 5, 10, 19, 23, 30, 34, 56, 65, 67, 86, 87, 100, 105, 112, 138
Southern African Development Community (SADC), 5
spectral refraction, 15
stakeholders, 108, 110, 126, 127, 130, 137
standardization, 29, 74
stockpiles, 90, 95
strict membership standards, 28
stricter controls, 89

T

tamper-proof containers, 26, 68
tariff agreements, 55
tariff reductions, 55
tax collection, 121, 127, 128

Index

tax evasion, 104, 126, 127
tax revenue, 75, 130
taxes, 41, 88, 89, 110, 115, 118, 119, 127, 130, 131, 133, 140
Taylor, Liberian President Charles, 8, 9, 19, 46, 47, 66
technical assistance, 15, 20, 51, 87, 104, 114, 126, 134, 135
Tempelsman, Maurice, 67, 71
terrorism, 6, 89, 92
terrorist groups, 6
Tiffany, 67, 70
Togo, 66
total diamond production, 32
tracking diamonds, 25
trade and customs policies, 26
trade documentation, 15, 26
trade groups, 11, 66
Trade in African Diamonds, 14, 17, 20, 21, 30, 57
trade in conflict diamonds, 67, 71, 81-84, 94, 97
trade in diamonds, 46, 54, 113
trade of conflict diamonds, vii
trade, 77, 78
travel ban, 78, 79
travel sanctions, 76
Treasury Department, 21, 69, 71

U

U.N. General Assembly (UNGA), 21, 24, 28, 61, 87
U.N. sanctions, 5, 21, 33, 34, 50, 82, 91
U.N. Security Council, 7, 8, 14, 19, 22, 29, 32, 33, 34, 41, 42, 49
U.S. Customs Service, 88
U.S. State Department, 7
U.S. Trade Representative, 23, 88
Ukraine, 66
UN peace-keepers, 123, 133
UN Security Council resolutions, 31
Union for the Total Independence of Angola (UNITA), 5, 7, 32-36, 62, 66, 67, 73-77, 85, 101
UNITA rebels, 66
United Arab Emirates, 89
United Kingdom (UK), 24, 26, 50, 51, 57, 77, 100, 101, 105, 113, 134, 135, 139
United Nations (UN), 1, 5, 22, 33, 35, 41, 49-51, 53, 55, 56, 60-62, 65, 73, 75-77, 83-87, 94, 95, 105, 111, 113, 132, 134, 135
United Nations General Assembly, 73, 83
United Nations Mission in Sierra Leone (UNAMSIL), 7, 46, 61, 77
United Nations reports, 65
United Nations sanctions, 86, 132
unmounted diamonds, 4
unofficial trade and production, 4
US Agency for International Development/Office of Transition Initiatives (USAID/OTI), 103-105, 107, 111, 126, 127, 130, 132, 133, 135, 138

V

value, vii, 4, 8, 41, 43, 44, 51, 81, 83, 89, 91, 104, 110, 112, 115, 116, 118, 119, 121, 125, 128-130

W

weight, 51, 91, 104, 116, 118
wholesale, 10, 18
World Bank, 3, 13, 45, 56, 61, 84, 93
World Diamond Congress, 16, 75, 106, 139
World Diamond Council (WDC), 12, 14, 16-18, 25-27, 42, 57, 66, 68-71, 87, 95, 106, 107, 125, 132, 134, 139
World Diamond Market, 3

World Federation of Diamond
  Bourses (WFDB), 16, 17, 52, 87,
  100
World Trade Organization (WTO),
  28, 30, 85, 88, 93, 94
World Vision, 65, 106

Z

Zaire, 66, 67
Zimbabwe, 66, 77
Zurel Brothers, 43, 118